Lecture Notes in Statistics

Edited by D. Brillinger, S. Fienberg, J. Gani, J. Hartigan, and K. Krickeberg

42

S. Kullback
J.C. Keegel
J.H. Kullback

Topics in Statistical Information Theory

Springer-Verlag
Berlin Heidelberg New York London Paris Tokyo

Authors

John C. Keegel
Department of Mathematics, University of District Columbia
Washington, D.C. 20008, USA

Joseph H. Kullback
Grumman-CTEC, Inc.
6862 Elm Street, McLean, VA 22101, USA

Solomon Kullback
Department of Statistics, George Washington University
Washington, D.C. 20052, USA

AMS Subject Classification (1980): 62-02

ISBN 3-540-96512-2 Springer-Verlag Berlin Heidelberg New York
ISBN 0-387-96512-2 Springer-Verlag New York Berlin Heidelberg

Printing and binding: Druckhaus Beltz, Hemsbach/Bergstr.
2147/3140-543210

PREFACE

The relevance of information theory to statistical theory and its applications to stochastic processes is a unifying influence in these TOPICS. The integral representation of discrimination information is presented in these TOPICS reviewing various approaches used in the literature, and is also developed herein using intrinsically information-theoretic methods. Log-likelihood ratios associated with various stochastic processes are computed by an application of minimum discrimination information estimates. Linear discriminant functionals are used in the information-theoretic analysis of a variety of stochastic processes.

Sections are numbered serially within each chapter, with a decimal notation for subsections. Equations, examples, theorems and lemmas, are numbered serially within each section with a decimal notation. The digits to the left of the decimal point represent the section and the digits to the right of the decimal point the serial number within the section. When reference is made to a section, equation, example, theorem or lemma within the same chapter only the section number or equation number, etc., is given. When the reference is to a section ,equation, etc., in a <u>different</u> chapter, then in addition to the section or equation etc., number, the chapter number is also given.

References to the bibliography are by the author's name followed by the year of publication in parentheses.

The transpose of a matrix is denoted by a prime; thus one-row matrices are denoted by primes as the transposes of one-column matrices (vectors). Clarification is included where considered necessary for notational usage.

It is our hope that the TABLE OF CONTENTS and the INDEX will assist in providing access to subject matter items of interest.

Acknowledgements

To the graduate students in Statistics at The George Washington University and Florida State University, who by their interest and searching questions led to improvements in the exposition of earlier versions of these topics, our appreciation.

These topics were prepared with the partial support of the Air Force Office of Scientific Research, Office of Aerospace Research, United States Air Force under Grant AFOSR-68-1513, and research activities of Grumman-CTEC.

The cooperation of The Department of Statistics, The George Washington University; The Department of Statistics, Florida State University; The Department of Mathematics, The University of the District of Columbia; and Grumman-CTEC is appreciated.

To Mrs. Lola S. Kullback for the first draft and Mary Crider and Carol Ann Quarto who prepared the final version and all others who have contributed, our sincere thanks.

The interest in the manuscript by Dr. Stefan Shrier of Grumman-CTEC contributed to its final preparation.

The consideration of the manuscript by Walter Kaufmann - Bühler, Mathematics Editor of Springer-Verlag Publishers, his favorable decision and editorial assistance are greatly appreciated.

Contents

Contents (Continued)

Contents (Continued)

I: INTRODUCTION

1. Purpose

The following collection of Topics in Statistical Information Theory are intended to complement, expand and extend subjects presented in Kullback(1959). Familiarity with Kullback(1959) or concurrent reference to it would be helpful to the reader of these topics, but not necessary. Whereas in Kullback(1959) attention was devoted to informational properties of statistics, that is, random variables, in these topics attention is focused on informational properties of sub-sigma-algebras of the fundamental probability space. Whereas in Kullback (1959) applications were to finite samples and vectors with a finite number of components, in these topics illustrations and examples will be given of applications to stochastic processes and random functions. This objective was indicated in the Preface to Kullback (1959). The notation in these topics differs somewhat from Kullback (1959) in that it follows Loève (1955).

2. Separable Sigma-Algebras

We shall assemble here certain results on separable sigma-algebras which we shall need (Loève (1955)).

2.1 A separable sigma-algebra is a sigma-algebra that is generated by (is minimal over) a countable class of sets.

2.2 The minimal sigma-algebra over the union of a countable class of separable sigma-algebras is also a separable sigma-algebra.

Let A_{ij}, j = 1, 2, . . . be the countable class of sets that generates the sigma-algebra $\mathbf{A}_i$, then the minimal sigma-algebra over the union of the $\mathbf{A}_i$'s is the sigma-algebra generated by the union of the A_{ij}'s which is also countable.

2.3 The Borel sigma-algebra on the real line is separable.

The Borel sigma-algebra on the real line is the minimal sigma-algebra over the class of all intervals of the form $(-\infty, S)$ where S is a real number. But every such interval can be written as

$$(-\infty, S) = \bigcup_{r<S} (-\infty, r),\ r \text{ a rational number.} \tag{2.1}$$

Since the set of rational numbers is countable, the Borel sigma-algebra on the real line is also the minimal sigma-algebra over the class of all intervals of the form $(-\infty, r)$ where r is a rational number and hence the Borel sigma-algebra is separable.

2.4 The inverse image of a separable sigma-algebra by a measurable transformation is a separable sigma-algebra.

This follows from the fact that the inverse image of the minimal sigma-algebra over a class of sets is the minimal sigma-algebra over the inverse image of the class of sets.

2.5 The sub-sigma-algebra induced by a random variable or a countable class of random variables is a separable sigma-algebra.

A random variable $X(\omega)$ on $(\Omega, \mathbf{A})$ is a measurable transformation taking values in $(R, \boldsymbol{\beta})$ where R is the real line and $\boldsymbol{\beta}$ is the Borel sigma-algebra on the real line, hence 2.3 and 2.4 imply the desired result.

2.6 A finite (countable) partition of a space Ω is a finite (countable) sequence of sets A_i such that

$$\sum A_i = \Omega,\quad A_i \bigcap A_j = \emptyset,\quad i \neq j. \tag{2.2}$$

If $\mathbf{A}$ is a sigma-algebra of subsets of Ω then the partition is measurable $\mathbf{A}$ if $A_i \in \mathbf{A}$ for all i. Let $\varepsilon = \{E_i\}$ be an $\mathbf{A}$-measurable partition. The $\mathbf{A}$-measurable partition

$\mathbf{D} = \{D_j\}$ is said to be a subpartition of ε or finer than the partition ε if each $D_j \in \mathbf{D}$ is such that $D_j \subset E_i \in \varepsilon$ and we denote this by $\mathbf{D} < \varepsilon$ or $\varepsilon > \mathbf{D}$.

A sequence of partitions $\{\varepsilon^n\}$ is said to be regular if each ε^n is a finite partition and

$$\varepsilon^1 > \varepsilon^2 > \varepsilon^3 \ldots . \tag{2.3}$$

2.7 Let $\underline{\varepsilon}^n$ denote the finite algebra generated by the partition ε^n, then corresponding to (2.3)

$$\underline{\varepsilon}^1 \subset \underline{\varepsilon}^2 \subset \underline{\varepsilon}^3 \subset \ldots \tag{2.4}$$

and $\bigcup_1^\infty \underline{\varepsilon}^n$ is an algebra. Let $\underline{\varepsilon}$ be the minimal sigma-algebra over $\bigcup_1^\infty \underline{\varepsilon}^n$, then $\underline{\varepsilon}$ is said to be generated by the regular sequence of partitions (2.3).

2.8 It is clear that if a sigma-algebra is generated by a regular sequence of partitions then it is separable. Conversely, every separable sigma-algebra $\mathbf{A}$ can be generated by a regular sequence of partitions.

Let $\{A_i\}$ be the sequence of sets that generates $\mathbf{A}$. Let ε^n be the class of all sets of the form $\bigcap_{i=1}^n A_i^*$ where A_i^* takes the value A_i or $A_i^c = \Omega - A_i$. The 2^n sets in the class ε^n comprise a partition of Ω, $\varepsilon^n > \varepsilon^{n+1}$ and the finite algebras they generate satisfy $\underline{\varepsilon}^n \subset \underline{\varepsilon}^{n+1}$. Also, $\varepsilon^n \subset \mathbf{A}$, $\underline{\varepsilon}^n \subset \mathbf{A}$ for all n, hence $\underline{\varepsilon}$ the minimal sigma-algebra over $\bigcup_1^\infty \underline{\varepsilon}^n$ is also contained in $\mathbf{A}$. But A_i is contained in one of the sets of $\underline{\varepsilon}^n$ for $i \leq n$, hence, $A_i \in \underline{\varepsilon}$ for all i which implies that $\mathbf{A} \subset \underline{\varepsilon}$. Thus, $\mathbf{A} = \underline{\varepsilon}$ and is generated by the regular sequence of partitions $\{\varepsilon^n\}$.

3. Probability Space

We shall list here certain notations which we shall use in these Topics.

We shall operate in the probability space $(\Omega, \mathbf{A}, P)$. Let $Z_t(\omega)$ and $Z(\omega)$ be non-negative random variables such that

$$\mu_t(A) = \int_A Z_t(\omega)\, dP(\omega), \;\; \mu(A) = \int_A Z(\omega)\, dP(\omega), \;\; A \in \mathbf{A} \tag{3.1}$$

are probability measures. We also write (3.1) in the Radon-Nikodym differential formalism as

$$d\mu_t = Z_t dP, \;\; d\mu = Z dP. \tag{3.2}$$

$Z_t(\omega)$ and $Z(\omega)$ may be considered as generalized probability densities. If we assume that μ_t is absolutely continuous with respect to μ, that is $\mu_t \ll \mu$, then

$$d\mu_t = W_t \, d\mu = W_t Z dP = Z_t dP, \;\; W_t = Z_t / Z \;\; a.s. \tag{3.3}$$

so that W_t is a likelihood ratio. We shall also require sequences of the generalized densities, corresponding probability measures, and likelihood ratios, that is,

$$d\mu_{tn} = Z_{tn} dP, \;\; d\mu_n = Z_n dP, \;\; n = 1, 2, \ldots \tag{3.4}$$

$$d\mu_{tn} = W_{tn} d\mu_n = W_{tn} Z_n dP = Z_{tn} dP, \;\; W_{tn} = Z_{tn} / Z_n \;\; a.s.$$

We shall have occasion to deal with the properties of relative conditional expectations as described in Loève (1955, p. 344).

Let β be a sub-sigma-algebra of the sigma-algebra $\mathbf{A}$, that is, $\beta \subset \mathbf{A}$. Corresponding to (3.2) and (3.3)

$$d\mu_\beta = E^\beta Z dP_\beta, \;\; d\mu_{t\beta} = E^\beta Z_t \, dP_\beta, \;\; d\mu_{t\beta} = E_Z^\beta W_t \, d\mu_\beta$$

$$E_Z^\beta W_t = E_Z^\beta \frac{Z_t}{Z} = (E^\beta Z \frac{Z_t}{Z}) / E^\beta Z = E^\beta Z_t / E^\beta Z, \tag{3.5}$$

where P_β is the restriction of P to β defined by $P_\beta(B) = P(B)$, $B \in \beta$, and

$$\mu_\beta(B) = \int_B Z dP = \int_B (E^\beta Z) dP_\beta , \; B \in \beta \tag{3.6}$$

$$\int_B (E_Z^\beta X) d\mu_\beta = \int_B X d\mu , \; B \in \beta \tag{3.7}$$

$$E^\beta Z X = E^\beta Z \, E_Z^\beta X . \tag{3.8}$$

II: DISCRIMINATION INFORMATION AND SUFFICIENCY IN A SUB-SIGMA-ALGEBRA

1. Discrimination Information - Integral Representation

We shall use the notation presented in section 3 of Chapter I without further mention. We shall use natural logarithms throughout these topics.

The result in Corollary 3.2, p. 16 Kullback (1959) suggests that the discrimination information in the sub-sigma-algebra $\beta \subset \mathbf{A}$ generated by the partition $\{B_i\}$, $i = 1, 2, \ldots, n$, $B_i \in \mathbf{A}$, $\sum_{i=1}^{n} B_i = \Omega$, be defined by

$$I(\beta : \mu_t, \mu) = \sum_{i=1}^{n} \mu_t(B_i) \ln \frac{\mu_t(B_i)}{\mu(B_i)}, \; B_i \in \beta \subset \mathbf{A}. \tag{1.1}$$

Because of the convexity of the function $x \ln (x/y)$ for non-negative x and y, and additivity of the measures for disjoint sets, for $A_1, A_2 \in \mathbf{A}$, $A_1 \cap A_2 = \varnothing$,

$$\begin{aligned} &\mu_t(A_1) \ln \frac{\mu_t(A_1)}{\mu(A_1)} + \mu_t(A_2) \ln \frac{\mu_t(A_2)}{\mu(A_2)} \\ &\geq (\mu_t(A_1) + \mu_t(A_2)) \ln \frac{\mu_t(A_1) + \mu_t(A_2)}{\mu(A_1) + \mu(A_2)} \\ &= \mu_t(A_1 + A_2) \ln \frac{\mu_t(A_1 + A_2)}{\mu(A_1 + A_2)} . \end{aligned} \tag{1.2}$$

The property in (1.2) suggests that the discrimination information in $\mathbf{A}$ be defined by (cf. Chiang Tse-Pei (1958), Dobrushin (1959, 1960), Gelfand et al (1956), Gelfand and Yaglom (1957), Ghurye (1969), Kallianpur (1960), Pérez (1957), Pinsker (1964))

$$\bar{I}(\mathbf{A}{:}\mu_t,\mu) = \mathrm{Sup}_{A_i \in \mathbf{A}} \sum \mu_t(A_i)\, ln\, \frac{\mu_t(A_i)}{\mu(A_i)} \tag{1.3}$$

where the sup is taken over all possible **A**-measurable finite partitions of Ω. For convenience hereafter we shall omit the μ_t and μ in $I(\beta{:}\mu_t,\mu)$ and $I(\mathbf{A}{:}\mu_t,\mu)$ unless needed for clarification.

If μ_t is not absolutely continuous with respect to μ, that is, there exist an $A \in \mathbf{A}$ such that $\mu(A) = 0$, $\mu_t(A) \neq 0$ then $\bar{I}(\mathbf{A}) = \infty$. Accordingly we shall assume that $\mu_t \ll \mu$. Note that $\bar{I}(\mathbf{A})$ may be infinite in this case also. $\mu_t \ll \mu$ is a necessary condition for $\bar{I}(\mathbf{A}) < \infty$. (See Kullback (1959, pp. 5, 6)).

It also seems reasonable to have directly defined the discrimination information in **A** by (cf. Kullback (1959, p. 5))

$$I(\mathbf{A}) = \int Z_t(\omega)\, ln\, \frac{Z_t(\omega)}{Z(\omega)}\, dP(\omega)\,. \tag{1.4}$$

The integral representation (1.4) may also be written as

$$I(\mathbf{A}) = \int Z_t\, ln\, \frac{Z_t}{Z}\, dP = \int W_t\, ln\, W_t\, d\mu = \int (ln\, W_t)\, d\mu_t = \int d\mu_t\, ln\, \frac{d\mu_t}{d\mu}\,. \tag{1.5}$$

Using the additivity of the integral and Jensen's inequality for convex functions

$$\int W_t\, ln\, W_t\, d\mu = \sum_{i=1}^{n} \int_{A_i} W_t\, ln\, W_t\, d\mu,\ A_i \in \mathbf{A},\ \sum_{i=1}^{n} A_i = \Omega, \tag{1.6}$$

$$\int_{A_i} W_t\, ln\, W_t\, d\mu \geq \int_{A_i} W_t\, d\mu\, ln\, \frac{\int_{A_i} W_t\, d\mu}{\int_{A_i} d\mu} = \mu_t(A_i)\, ln\, \frac{\mu_t(A_i)}{\mu(A_i)}\,, \tag{1.7}$$

$$\int W_t \ln W_t \, d\mu \geq \sum_{i=1}^{n} \mu_t(A_i) \ln \frac{\mu_t(A_i)}{\mu(A_i)}, \tag{1.8}$$

hence

$$I(\mathbf{A}) = \int W_t \ln W_t \, d\mu \geq \operatorname{Sup}_{A_i \in \mathbf{A}} \sum_{i=1}^{n} \mu_t(A_i) \ln \frac{\mu_t(A_i)}{\mu(A_i)} = \bar{I}(\mathbf{A}), \tag{1.9}$$

where the sup is taken over all possible **A**-measurable finite partitions of Ω. A proof of the reverse of the inequality in (1.9) may be obtained following the method in Pinsker (1964, pp. 24-25) or Ghurye (1968). We shall not pursue this matter any further at this point but state the integral representation theorem.

<u>Theorem 1.1</u> $\bar{I}(\mathbf{A}) = I(\mathbf{A})$, that is (1.3) and (1.5) define the same value of the discrimination information.

Proofs of this theorem have involved martingale theory (e.g., Kallíanpur (1960)) or the use of the convexity property in conjunction with the Darboux-Young approach to the integral (Loève (1955, p. 143 Ex. 29) in Ghurye (1968). The proof to be presented in section 3 of Chapter III in connection with other results is believed to be intrinsically more information-theoretic in nature. For other approaches see Dobrushin (1959), Gelfand and Yaglom (1957), Pérez (1957), Pinsker (1964), and Guiasu (1968).

Note that if in the probability space $(\Omega, \mathbf{A}, P)$, **A** is generated by a finite partition $\{A_i\}$, then (3.1) in Chapter I is

$$\mu_t(A_i) = Z_t(\omega) P(A_i), \; \mu(A_i) = Z(\omega) P(A_i), \; \omega \in A_i \tag{1.10}$$

and both (1.3) and (1.5) yield

$$\bar{I}(\mathbf{A}) = \sum_{i=1}^{n} \mu_t(A_i) \ln \frac{\mu_t(A_i)}{\mu(A_i)} = I(\mathbf{A}). \tag{1.11}$$

We remark that instead of starting with the probability space $(\Omega, \mathbf{A}, P)$ we could have started with the measure space $(\Omega, \mathbf{A}, \lambda)$ where λ is a sigma-finite measure on **A**. We assume the existence of the non-negative **A**-measurable function $X(\omega)$ such that

$$P(A) = \int_A X(\omega)\, d\lambda\,(\omega),\ A \in \mathbf{A} \tag{1.12}$$

is a probability measure on **A**. We then have (see Kullback (1959), p. 5))

$$d\mu_t = Z_t\, dP = Z_t X d\lambda = f_t d\lambda,\ d\mu = Z dP = ZX d\lambda = f d\lambda\,[\lambda] \tag{1.13}$$

$$\int Z_t \ln \frac{Z_t}{Z}\, dP = \int Z_t \ln \frac{Z_t}{Z}\, X d\lambda = \int f_t \ln \frac{f_t}{f}\, d\lambda\,. \tag{1.14}$$

2. Discrimination Information in a Sub-Sigma-Algebra

The following discussion essentially extends the presentation in Kullback (1959, pp. 1-78). We shall use the notation and in particular properties developed in Loève (1955), Chapter VII, Conditioning, pp. 337 ff. In particular we shall use the fact that Jensen's inequality holds a.s. for conditional expectation also. Let β be a sub-sigma-algebra of the sigma-algebra **A**. The basic inequality is that if g is a convex function and EX is finite, then

$$E\{g(X)\} \geq g(EX). \tag{2.1}$$

In the conditional form, we have

$$E^{\beta}\{g(X)\} \geq g(E^{\beta} X)\ a.s. \tag{2.2}$$

Using the definition of conditional expectation, (2.2) and (1.4),

$$I(\mathbf{A}) = \int Z_t \ln \frac{Z_t}{Z}\, dP = \int E^\beta (Z_t \ln \frac{Z_t}{Z})\, dP_\beta \geq \int E^\beta Z_t \ln \frac{E^\beta Z_t}{E^\beta Z}\, dP_\beta \,. \tag{2.3}$$

We define the right-hand side of (2.3) as the discrimination information in the sub-sigma-algebra β and note that

$$\begin{aligned} I(\beta) &= \int E^\beta Z_t \ln \frac{E^\beta Z_t}{E^\beta Z}\, dP_\beta = \int E_Z^\beta W_t \ln E_Z^\beta W_t\, d\mu_\beta \\ &= \int (\ln E_Z^\beta W_t)\, d\mu_{t\beta} = \int d\mu_{t\beta} \ln \frac{d\mu_{t\beta}}{d\mu_\beta} \\ &= \operatorname{Sup} \sum_{k=1}^{n} \mu_t(B_k) \ln \frac{\mu_t(B_k)}{\mu(B_k)} \end{aligned} \tag{2.4}$$

where the Sup is taken over all finite β-measurable partitions of Ω (see (2.15) and section 1 of this chapter). We can now state:

Theorem 2.1 If β is a sub-sigma-algebra of **A** then

$$I(\mathbf{A}) \geq I(\beta)\,. \tag{2.5}$$

Note that the coarsest possible sub-sigma-algebra is that generated by $(\varnothing, \Omega)$ and denoting it by β_o, $E^{\beta_o} Z_t = EZ_t = 1$, $E^{\beta_o} Z = EZ = 1$ and

$$I(\beta_o) = 0. \tag{2.6}$$

Theorem 2.2 $I(\mathbf{A}) \geq 0$, with equality if and only if $W_t = Z_t/Z = 1$ a.s. (See Kullback (1959), Theorem 3.1, p. 14.)

From (1.5) and (2.4), using the result that

$$\int W_t \, ln \, E_Z^{\beta} W_t d\mu = \int E_Z^{\beta}(W_t \, ln \, E_Z^{\beta} W_t) \, d\mu_{\beta} = \int E_Z^{\beta} W_t \, ln \, E_Z^{\beta} W_t \, d\mu_{\beta} \tag{2.7}$$

we have

$$\begin{aligned} I(\mathbf{A}) - I(\beta) &= \int W_t \, ln \, W_t \, d\mu - \int E_Z^{\beta} W_t \, ln \, E_Z^{\beta} W_t \, d\mu_{\beta} \\ &= \int W_t \, ln \, (W_t / E_Z^{\beta} W_t) \, d\mu \\ &= \int \frac{Z_t}{Z} \, ln \, \frac{Z_t/Z}{E^{\beta} Z_t / E^{\beta} Z} \, d\mu \, . \end{aligned} \tag{2.8}$$

Since

$$\begin{aligned} &\int W_t \, d\mu = \int \frac{Z_t}{Z} \, d\mu = \int Z_t \, dP = 1, \\ &\int E_Z^{\beta} W_t \, d\mu_{\beta} = \int (E^{\beta} Z_t / E^{\beta} Z) \, d\mu_{\beta} \\ &= \int E^{\beta} Z_t \, dP_{\beta} = \int Z_t \, dP = 1, \end{aligned} \tag{2.9}$$

the right-hand side of (2.8) is a discrimination information value, and as such non-negative. Let us define using (2.8) above and (3.2) and (3.5) of Chapter I

$$I(\mathbf{A}|\beta) = \int W_t \, ln \, (W_t / E_Z^{\beta} W_t) \, d\mu = \int Z_t \, ln \, \frac{Z_t / E^{\beta} Z_t}{Z / E^{\beta} Z} \, dP = \int d\mu_t \, ln \, \frac{d\mu_t / d\mu_{t\beta}}{d\mu / d\mu_{\beta}} \tag{2.10}$$

that is, $I(\mathbf{A}|\beta)$ is the conditional discrimination information in **A** given β and hence

<u>Theorem 2.3.</u> If β is a sub-sigma-algebra of **A**

$$I(\mathbf{A}) = I(\beta) + I(\mathbf{A}|\beta) \, . \tag{2.11}$$

<u>Theorem 2.4</u> If β is a sub-sigma-algebra of **A**, then

$$I(\mathbf{A}) = I(\beta) \tag{2.12}$$

if and only if $I(\mathbf{A}|\beta) = 0$, that is, if and only if

$$Z_t/Z = E^{\beta}Z_t/E^{\beta}Z \quad a.s. \tag{2.13}$$

Proof. Apply Theorem 2.2 and (2.10).

If β is a sub-sigma-algebra of $\mathbf{A}$ and satisfies Theorem 2.4 then we say that β is a sufficient sub-sigma-algebra for $\mathbf{A}$. (See Loève (1955, p. 346), Kullback (1959, pp. 18-22).)

If β is generated by a finite partition, then on a non-null atom $B \in \beta$, the conditional expected value $E^{\beta}X$ is a constant and its value is

$$E^{\beta}X = \frac{1}{P(B)} \int_B X dP, \ \omega \in B. \tag{2.14}$$

Thus

$$\begin{aligned} I(\beta) &= \int E^{\beta}Z_t \ln \frac{E^{\beta}Z_t}{E^{\beta}Z} dP_{\beta} = \sum_i \int_{B_i} E^{\beta}Z_t \ln \frac{E^{\beta}Z_t}{E^{\beta}Z} dP_{\beta} \\ &= \sum_i P(B_i) \frac{1}{P(B_i)} \int_{B_i} Z_t \, dP \ln \frac{\int_{B_i} Z_t \, dP}{\int_{B_i} Z \, dP} \\ &= \sum_i \mu_t(B_i) \ln \frac{\mu_t(B_i)}{\mu(B_i)} . \end{aligned} \tag{2.15}$$

Note that in this case

$$\frac{Z}{E^{\beta}Z} = Z/(\frac{1}{P(B)} \int_B Z \, dP) = Z/(\mu(B)/P(B)), \ \omega \in B \tag{2.16}$$

and

$$
\begin{aligned}
I(\mathbf{A}|\beta) &= \int Z_t \, ln \, \frac{Z_t/E^\beta Z_t}{Z/E^\beta Z} \, dP \\
&= \sum_i \int_{B_i} Z_t \, ln \, \frac{Z_t/E^\beta Z_t}{Z/E^\beta Z} \, dP \\
&= \sum_i \int_{B_i} Z_t \, ln \, \frac{Z_t P(B_i) \mu(B_i)}{\mu_t(B_i) Z P(B_i)} \, dP \qquad (2.17) \\
&= \sum_i \mu_t(B_i) \int_{B_i} \frac{Z_t}{\mu_t(B_i)} \, ln \, \frac{Z_t/\mu_t(B_i)}{Z/\mu(B_i)} \, dP \\
&= \sum_i \mu_t(B_i) I(\mathbf{A}|B_i)
\end{aligned}
$$

where

$$
\begin{aligned}
I(\mathbf{A}|B_i) &= \int_{B_i} \frac{Z_t}{\mu_t(B_i)} \, ln \, \frac{Z_t/\mu_t(B_i)}{Z/\mu(B_i)} \, dP \\
&= \int_{B_i} \frac{d\mu_t}{\mu_t(B_i)} \, ln \, \frac{d\mu_t/\mu_t(B_i)}{d\mu/\mu(B_i)} \\
&= \frac{1}{\mu_t(B_i)} \int_{B_i} Z_t \, ln \, \frac{Z_t}{Z} \, dP - ln \, \frac{\mu_t(B_i)}{\mu(B_i)} \qquad (2.18) \\
&= \frac{1}{\mu_t(B_i)} \int_{B_i} W_t \, ln \, W_t d\mu - ln \, \frac{\mu_t(B_i)}{\mu(B_i)} \, .
\end{aligned}
$$

Note that $I(\mathbf{A}|B_i)$ is a discrimination information, and

$$I(\mathbf{A}|B_i) \geq 0 \tag{2.19}$$

with equality if and only if

$$Z_t/Z = \mu_t(B_i)/\mu(B_i),\ \omega \in B_i, \tag{2.20}$$

(see Kullback (1959), Corollary 3.1, p. 15.)

To obtain another representation for $I(\mathbf{A}|\,\beta)$ when β is a separable sub-sigma-algebra of $\mathbf{A}$ we proceed as follows. It follows from (2.18) that

$$\begin{aligned}\sum_i \mu_t(B_i) I(\mathbf{A}|B_i) &= \sum_i \int_{B_i} Z_t \ln \frac{Z_t}{Z}\, dP - \sum_i \mu_t(B_i) \ln \frac{\mu_t(B_i)}{\mu(B_i)} \\ &= \int Z_t \ln \frac{Z_t}{Z}\, dP - \sum_i \mu_t(B_i) \ln \frac{\mu_t(B_i)}{\mu(B_i)} \\ &= I(\mathbf{A}) - \sum_i \mu_t(B_i) \ln \frac{\mu_t(B_i)}{\mu(B_i)}\,.\end{aligned} \tag{2.21}$$

Now let $\underline{\pi}(\beta)$ denote the class of finite β-measurable partitions of Ω, then (see Ghurye (1968))

$$\begin{aligned}\inf_{\underline{\pi}(\beta)} \sum_i \mu_t(B_i) I(\mathbf{A}|B_i) &= \inf_{\underline{\pi}(\beta)} \{I(\mathbf{A}) - \sum_i \mu_t(B_i) \ln \frac{\mu_t(B_i)}{\mu(B_i)}\} \\ &= I(\mathbf{A}) - \sup_{\underline{\pi}(\beta)} \sum_i \mu_t(B_i) \ln \frac{\mu_t(B_i)}{\mu(B_i)} \\ &= I(\mathbf{A}) - I(\beta) = I(\mathbf{A}|\beta)\,.\end{aligned} \tag{2.22}$$

From (2.22) we note that $I(\beta|\beta) = 0$.

3. Sufficiency

Let W_t be defined as in (3.3) of Chapter I, and denote by β_{W_t} the class of sets $W_t^{-1}(B)$ where B ranges over linear Borel sets, that is

$$W_t^{-1}(B) = \{\omega : W_t(\omega) \in B\}. \tag{3.1}$$

β_{W_t} is the minimal sigma-algebra with respect to which W_t is measurable and β_{W_t} is a separable sub-sigma-algebra of **A**. For convenience we shall hereafter denote β_{W_t} by β_t. We shall now show that β_t is a sufficient sub-sigma-algebra for **A**.

Theorem 3.1 β_t is a sufficient sub-sigma-algebra for **A**, that is, $I(\mathbf{A}) = I(\beta_t)$.

Proof.

$$\begin{aligned} I(\mathbf{A}) &= \int W_t \ln W_t \, d\mu = \int E_Z^{\beta_t}(W_t \ln W_t) \, d\mu_{\beta_t} \\ &\geq \int E_Z^{\beta_t} W_t \ln E_Z^{\beta_t} W_t \, d\mu_{\beta_t} = I(\beta_t) = \int W_t \ln W_t \, d\mu = I(\mathbf{A}) \end{aligned} \tag{3.2}$$

where we have used the fact that since W_t is measurable β_t then $E_Z^{\beta_t} W_t = W_t [\mu]$.

Note that using (3.5) of Chapter I, $E^{\beta_t} Z_t / E^{\beta_t} Z = Z_t / Z$, which is the necessary and sufficient condition (2.13) for Theorem 2.4.

In order to extend the concept of sufficiency to a family of distributions, we now consider the family $\{Z_t, t \in T\}$ such that (3.3) of Chapter I holds for all $t \in T$. We can now apply Theorem 2.4 to state

Theorem 3.2. The sub-sigma-algebra β is sufficient for the family $\{Z_t\}$ if and only if, there exists a Z such that every $Z_t = W_t Z$ a.s. and every W_t *is* β-measurable; then every $W_t = Z_t/Z = E^\beta Z_t / E^\beta Z$ up to a μ_β-equivalence, that is, except for sets of μ_β-measure zero.

Note that the statement in Theorem 3.2 implies that the family of probability measures $\{\mu_t, t \in T\}$ is dominated by μ. If we now let β_t denote the sub-sigma-algebra induced by the family $\{W_t, t \in T\}$, that is, every W_t is β_t-measurable, then β_t is the minimal or coarsest sigma-algebra sufficient with Z.

It may be shown that for the family of measures in Theorem 3.2, there exists a Z_o such that (see Loève (1955), p. 141)

$$\mu_o A = 0 \Leftrightarrow \mu_t A = 0 \text{ or } Z_o = 0 \text{ on } A \Leftrightarrow Z_t = 0 \text{ on } A\,[P], \text{ all } t. \tag{3.3}$$

In this case the family of measures $\{\mu_o, \mu_t\}$, is said to be homogeneous. Because of (3.3) every $E_Z^\beta X$ common to all the equivalence classes $E_{Z_t}^\beta X$ also belongs to the equivalence class of $E_{Z_o}^\beta X$, hence every sub–sigma–algebra β sufficient with Z is also sufficient with Z_o. Accordingly the least fine sufficient sub-sigma-algebra with Z_o is the coarsest one. We may now state Theorem 3.3.

<u>Theorem 3.3.</u> Write every Z_t in the form $Z_t = W_t Z_o$ a.s., with Z_o such that every $Z_t = 0$ *a.s.* $\Rightarrow Z_o = 0$ a.s. Then the coarsest (minimal) sufficient sub-sigma-algebra for the family $\{Z_t\}$ is the sub-sigma-algebra β_t induced by the family $\{W_t\}$. (See Loève (1955), pp. 344-347, Kullback (1959), pp. 18-22; 28-31; 43-45.)

We shall now show that the property of sufficiency is not affected by truncation or by selection of observations (Kullback (1959), pp. 21-22). Suppose that $X(\omega)$ is a non-negative function such that XZ_t, XZ are integrable, where Z_t and Z are defined in (3.2) and (3.3) of Chapter I, and let β be a sub-sigma-algebra of **A**. We have

$$\int XZ_t \ln \frac{XZ_t}{XZ}\, dP \geq \int E^\beta XZ_t \ln \frac{E^\beta XZ_t}{E^\beta XZ}\, dP_\beta \tag{3.4}$$

with equality if and only if

$$\frac{XZ_t}{XZ} = \frac{Z_t}{Z} = \frac{E^{\beta}XZ_t}{E^{\beta}XZ} = \frac{E^{\beta}_{Z_t} X E^{\beta}_{Z_t}}{E^{\beta}_{Z} X E^{\beta} Z} \quad a.s \tag{3.5}$$

that is if and only if β is a sufficient sub-sigma-algebra (Loève (1955), pp. 345-346). Suppose X is a β-measurable function, then

$$E^{\beta}XZ_t = XE^{\beta}Z_t, \; E^{\beta}XZ = XE^{\beta}Z \quad a.s. \tag{3.6}$$

and (3.4) becomes

$$\int XZ_t \, ln \, \frac{Z_t}{Z} \, dP \geq \int XE^{\beta}Z_t \, ln \, \frac{E^{\beta}Z_t}{E^{\beta}Z} \, dP_{\beta} \tag{3.7}$$

with equality if and only if β is a sufficient sub-sigma-algebra. In particular, suppose $X(\omega) = I_B(\omega), B \in \beta$, then (3.7) becomes

$$\int_B Z_t \, ln \, \frac{Z_t}{Z} \, dP \geq \int_B E^{\beta}Z_t \, ln \, \frac{E^{\beta}Z_t}{E^{\beta}Z} \, dP, \; B \in \beta \tag{3.8}$$

with equality if and only if β is a sufficient sub-sigma-algebra.

Note that if β in (3.8) is generated by a finite partition, then as in (2.16) $E\beta Z = \mu(B)/P(B)$ and $E\beta Z_t = \mu_t(B)/P(B)$ for $\omega \in B$ a non-null atom, and the right side of (3.8) becomes

$$\int_B \frac{\mu_t(B)}{P(B)} \, ln \, \frac{\mu_t(B)}{\mu(B)} \, dP_{\beta} = \mu_t(B) \, ln \, \frac{\mu_t(B)}{\mu(B)}, \, B \in \beta \tag{3.9}$$

and using (2.18) we may write

$$\int_B Z_t \ln \frac{Z_t}{Z}\, dP = \int_B E^{\beta} Z_t \ln \frac{E^{\beta} Z_t}{E^{\beta} Z}\, dP_{\beta} + \mu_t(B) I(\mathbf{A}|B) \tag{3.10}$$

and in this case $I(\beta|B) = 0$.

If β is a separable sub-sigma-algebra of $\mathbf{A}$, then using the notation as in (2.18), that is,

$$I(\mathbf{A}|B) = \frac{1}{\mu_t(B)} \int_B Z_t \ln \frac{Z_t}{Z} P - \ln \frac{\mu_t(B)}{\mu(B)}, \; B \in \beta$$

$$I(\beta|B) = \frac{1}{\mu_t(B)} \int_B E^{\beta} Z_t \ln \frac{E^{\beta} Z_t}{E^{\beta} Z}\, dP_{\beta} - \ln \frac{\mu_{t\beta}(B)}{\mu_{\beta}(B)}, \; B \in \beta, \tag{3.11}$$

it is seen that (3.8) may also be written as (cf. (4.13) in Chapter III)

$$I(\mathbf{A}|B) \geq I(\beta|B), \; B \in \beta \tag{3.12}$$

remembering that $\mu_t(B) = \mu_{t\beta}(B)$, $\mu(B) = \mu_{\beta}(B)$, $B \in \beta$. Note that (3.11) or (3.12) implies $I(\mathbf{A}|\beta) \geq I(\beta|\beta) = 0$. The β-measurable random variable $X(\omega)$ in (3.7) defines a truncation or selection process according as

$$X(\omega) = I_B(\omega), \; B \in \beta \;\; or \;\; \int_B X(\omega) dP = \nu(B), \; B \in \beta \tag{3.13}$$

and $\nu(B)$ is a probability measure on β.

4. Minimum Discrimination Information Theorem

This section is essentially the same as Kullback and Khairat (1966), except that we shall consider a measure space rather than a probability measure space.

Let Ω be a space of points ω, $\mathbf{A}$ a sigma-algebra of sets of Ω and λ a sigma-finite measure on $\mathbf{A}$. Let $T(\omega)$ be a real-valued $\mathbf{A}$-measurable function such that

$$M = \int exp\,(T\,(\omega)\,)\,d\lambda(\omega) < \infty \tag{4.1}$$

and let a probability measure P^* be defined by

$$P^*(A) = \int_A \,(exp\,T(\omega)/M)\,d\lambda(\omega),\ \ A \in \mathbf{A}. \tag{4.2}$$

Suppose that $T(\omega)$ is P^*-integrable, and let

$$\Theta = \int T(\omega)\,dP^*\,(\omega). \tag{4.3}$$

Now let P be an arbitrary probability measure on $\mathbf{A}$. If $P \ll \lambda$ define

$$I(P{:}\lambda\,) = \int \,[ln\,\frac{dP}{d\lambda}\,]\,dP \tag{4.4}$$

otherwise define $I(P{:}\lambda) = \infty$. It is clear that $I(P^*{:}\lambda)$ is finite; in fact, from (4.2) and (4.3)

$$I(P^*{:}\lambda) = \int \,(T(\omega) - ln\,M)dP^* = \Theta - ln\,M. \tag{4.5}$$

<u>Theorem 4.1.</u> If P is a probability measure on $\mathbf{A}$ such that $T(\omega)$ is P-integrable and

$$\int T(\omega)dP = \Theta \tag{4.6}$$

then

$$I(P{:}\lambda) \geq I(P^*{:}\lambda) = \Theta - ln\,M \tag{4.7}$$

with equality if and only if $P = P^*$ on $\mathbf{A}$.

Proof: If $I(P{:}\lambda) = \infty$ there is nothing to prove. Suppose then that $I(P{:}\lambda) < \infty$. In this case $P \ll \lambda$, and we write $f(\omega) = dP/d\lambda$. Then

$$I(P{:}\lambda) = \int f(\omega)\, ln\, f(\omega)\, d\lambda \tag{4.8}$$

and

$$I(P^*{:}\lambda) = \int f^*(\omega)\, ln\, f^*(\omega)\, d\lambda \tag{4.9}$$

where $f^*(\omega) = dP^*/d\lambda = (exp\, T(\omega))/M$ by (4.2). In view of (4.3), (4.6) and $f^*(\omega)$ as just defined, we have that

$$\int f(\omega)\, ln\, f^*(\omega)\, d\lambda = \int f^*(\omega)\, ln\, f^*(\omega)\, d\lambda. \tag{4.10}$$

We see from (4.8) - (4.10) that it suffices to show that the left-hand side of (4.10) cannot exceed the right-hand side of (4.8) that is, that

$$\int f(\omega)\, ln\, \frac{f^*(\omega)}{f(\omega)}\, d\lambda \leq 0. \tag{4.11}$$

Since $\ln Z \leq Z - 1$ for all $Z \geq 0$ with equality if and only if $Z = 1$, and since

$$\int f(\omega) d\lambda = \int f^*(\omega) d\lambda = 1 \tag{4.12}$$

it follows that (4.11) holds with equality if and only if $P = P^*$ on $\mathbf{A}$. The same conclusion may also be reached by using Jensen's inequality on the convex integrand

$$\int f(\omega)\, ln\, \frac{f(\omega)}{f^*(\omega)}\, d\lambda \geq \int f(\omega)d\lambda\, ln\, \frac{\int f(\omega)d\lambda}{\int f^*(\omega)d\lambda} = 1\, ln\, \frac{1}{1} = 0 \qquad (4.13)$$

with equality if and only if $f(\omega) = f^*(\omega)$ $[\lambda]$. We remark that using (4.10) we have the Pythagorean type relation

$$I(P{:}\lambda) = I(P^*{:}\lambda) + I(P{:}P^*) \qquad (4.14)$$

with $I(P{:}P^*) = \int f(\omega)\, ln\, (f(\omega)/f^*(\omega))\, d\lambda = \int ln\, (dP/dP^*)\, dP$.

Note that since λ is a sigma-finite measure $I(P{:}\lambda)$ is not necessarily non-negative.

<u>Example 4.1</u> Let $\Omega = \{0, 1, 2, \ldots\}$, **A** is the collection of all sets of non-negative integers and λ is counting measure. Let $T(\omega) = t\omega$, then

$$M = \int e^{t\omega} d\lambda = \sum_{\omega=0}^{\infty} e^{t\omega} = 1 + e^t + e^{2t} + e^{3t} + \ldots = 1/(1 - e^t), \qquad (4.15)$$

provided $e^t < 1$. It is found that

$$f^*(\omega) = e^{t\omega}(1 - e^t),\ t\theta = \int t\omega e^{t\omega}(1 - e^t)\, d\lambda = \theta$$

$$\theta = \frac{d}{dt}\, ln\, M = \frac{e^t}{1 - e^t} \qquad (4.16)$$

$$e^t = \theta/(1 + \theta),\ 1 - e^t = 1/(1 + \theta).$$

Hence in terms of the parameter $\theta > 0$

$$f^*(\omega) = \theta^{\omega}/(1 + \theta)^{1 + \omega} \qquad (4.17)$$

$$I(P^*:\lambda) = \theta\, ln\, \frac{\theta}{1+\theta} + ln\, \frac{1}{1+\theta} = \theta\, ln\, \theta - (\theta + 1)\, ln\, (\theta + 1) \tag{4.18}$$

$$\int f^*(\omega)\, ln\, f^*(\omega) d\lambda = \sum_{n=0}^{\infty} P^*(n)\, ln\, P^*(n). \tag{4.19}$$

The minimum discrimination information here corresponds to the maximum entropy where the entropy is defined by

$$H(P^*:\lambda) = -\, I(P^*:\lambda)\,. \tag{4.20}$$

Thus, given the mean value, the distribution on Ω with maximum entropy is the geometric distribution and the value of the maximum entropy is

$$H(P^*:\lambda) = (\theta + 1)\, ln\, (\theta + 1) - \theta\, ln\, \theta,\ \theta > 0\,. \tag{4.21}$$

In particular for $\theta = 1$, $f^*(\omega) = \frac{1}{2}(\frac{1}{2})^{\omega}$, $H = 2\, ln\, 2$.

Example 4.2 Let $\Omega = 0 \le \omega < \infty$, **A** the sigma-algebra of Borel sets and λ be Lebesgue measure.

Let $T(\omega) = -\, t\omega$, $t > 0$, then

$$M = \int_0^{\infty} e^{-t\omega}\, d\omega = \frac{1}{t}\,,\ f^*(\omega) = te^{-t\omega},\ + t\theta = \int_0^{\infty} t\omega e^{-t\omega}\, td\omega$$
$$= \theta \tag{4.22}$$
$$\theta = -\, \frac{d}{dt}\, ln\, M = \frac{d}{dt}\, ln\, t = \frac{1}{t}\,.$$

Hence in terms of the parameter θ

$$f^*(\omega) = e^{-\omega/\theta}\, \frac{1}{\theta},\ \int_0^{\infty} f^*(\omega)\, ln\, f^*(\omega) d\omega = ln\, \theta - 1 = -\, ln\, e\theta \tag{4.23}$$

(Kullback (1959), problem 8.27, p. 33). In this example, which is the continuous analogue of example 4.1 the f* distribution is an exponential.

5. Analysis of Information

We shall say that a random variable is a sufficient statistic if the sub-sigma-algebra induced by the random variable is a sufficient sub-sigma-algebra. We shall now consider an exponential type family and its sufficient statistics. Let $T(\omega) = (T_1(\omega), T_2(\omega), \ldots, T_n(\omega))$ be an **A**-measurable function and denote

$$tT(\omega) = t_1 T_1(\omega) + t_2 T_2(\omega) + \ldots + t_n T_n(\omega) \tag{5.1}$$

where $t = (t_1, t_2, \ldots, t_n)$ is a point in real R_n. We shall assume that $T(\omega)$ is such that

$$\int (\exp t\, T(\omega))\, Z(\omega)\, dP(\omega) = M(t) < \infty \tag{5.2}$$

for all t in some open convex region in R_n.

Now let $Y(\omega)$ be a generalized density belonging to the family **Y** of distributions satisfying the linearly independent constraints

$$\int T(\omega)\, Y(\omega)\, dP(\omega) = \theta \tag{5.3}$$

where $T(\omega)$ is defined in (5.1) and $\theta = (\theta_1, \theta_2, \ldots, \theta_n)$ with

$$\int T_i(\omega)\, Y(\omega)\, dP(\omega) = \theta_i,\ i = 1, 2, \ldots, n. \tag{5.4}$$

For given $Z(\omega)$, the minimum value of

$$I(Y:Z) = \int Y\omega)\, \ell n \frac{Y(\omega)}{Z(\omega)}\, dP(\omega) \tag{5.5}$$

for $Y(\omega) \in \mathbf{Y}$ occurs for (Kullback (1959, p. 38), Kullback and Khairat (1966)), (see theorem 4.1)

$$Y^*(\omega) = Z_t(\omega) = \frac{(\exp tT(\omega))}{M(t)} Z(\omega) \ a.s., \tag{5.6}$$

with $t = (t_1, t_2, \ldots, t_n)$ satisfying

$$\theta_i = \frac{\partial}{\partial t_i} \ell n\, M(t), \ i = 1, 2, \ldots, n \tag{5.7}$$

and

$$\int T(\omega) Y^*(\omega)\, dP(\omega) = \int T(\omega) Z_t(\omega)\, dP(\omega) = \theta. \tag{5.8}$$

The components of $t = (t_1, t_2, \ldots, t_n)$ are essentially Lagrange undetermined multipliers. Since $\ell n(Z_t/Z)$ is linear in $T(\omega)$, it may be shown that (a Pythagorean property, see (4.14))

$$I(Y{:}Z) = I(Y^*{:}Z) + I(Y{:}Y^*), \ Y \in \mathbf{Y}, \tag{5.9}$$

and

$$I(Y^*{:}Z) = t\theta - \ell n M(t), \tag{5.10}$$

where $t\theta$ represents $t_1\theta_1 + t_2\theta_2 + \ldots + t_n\theta_n$. We say that the family of probability measures

$$\mu_t(A) = \int_A \frac{\exp(t\ T(\omega))}{M(t)} Z(\omega)\, dP(\omega), A \in \mathbf{A} \tag{5.11}$$

that is

$$W_t = (\exp t\,T(\omega))/M(t),\; Z_t = \frac{(\exp t\,T(\omega))}{M(t)}\,Z,\ a.s. \tag{5.12}$$

is an exponential type family induced by $Z(\omega)$. Since

$$\frac{\partial}{\partial t_i}\,\ell n\,Y^*(\omega) = T_i(\omega) - \frac{\partial}{\partial t_i}\,\ell n\,M(t) = T_i(\omega) - \theta_i \tag{5.13}$$

we observe that the $T_i(\omega) = \hat{\theta}_i$ are the maximum likelihood estimators of the θ_i in $Y^*(\omega)$. Note that the family of measures $\{\mu_t\}$ is dominated by μ. If β is the sub-sigma-algebra induced by the measurable function $T(\omega)$, then

$$E_Z^\beta W_t = E_Z^\beta[(\exp t\,T(\omega))/M(t)] = (\exp t\,T(\omega))/M(t)\,[\mu_\beta] \tag{5.14}$$

that is, the necessary and sufficient condition of Theorem 3.2, and β is the minimal sigma-algebra sufficient with Z. In this case Z plays the same role as Z_o of Theorem 3.3, in fact, Z or μ is the member of the family corresponding to $t = 0$, and β is the coarsest (minimal) sub-sigma-algebra sufficient for the exponential family $\{\mu_t\}$, that is $T(\omega)$ is a function of any other sufficient statistic (Kullback (1959), pp. 45-55). If we write

$$W = Y/Z,\; W_t = Z_t/Z,\; W/W_t = Y/Z_t, \tag{5.15}$$

then in view of (5.6) we may write (5.9) as

$$I(Y{:}Z) = I(Z_t{:}Z) + I(Y{:}Z_t), \tag{5.16}$$

or

$$\int W\,ln\,W\,d\mu = \int W_t\,ln\,W_t\,d\mu + \int \frac{W}{W_t}\,ln\,\frac{W}{W_t}\,d\mu_t. \tag{5.17}$$

If, as in (5.14), β is the sub-sigma-algebra induced by the measurable function $T(\omega)$, then recalling (2.10) we see that (5.17) may be written as

$$I(\mathbf{A}) = I(\beta) + I(\mathbf{A}|\beta). \tag{5.18}$$

As a matter of fact a more detailed analysis of the information values may be made. Define $\mathbf{Y}_r$ as the family of distributions satisfying the linearly independent constraints

$$\int T_i(\omega)\, Y(\omega)\, dP(\omega) = \theta_i, \ i = 1, 2, \ldots, r \tag{5.19}$$

so that

$$\mathbf{Y}_1 \supset \mathbf{Y}_2 \supset \cdot \quad \cdot \quad \cdot \supset \mathbf{Y}_n = \mathbf{Y}. \tag{5.20}$$

It may be shown that

$$\begin{aligned} I(Y{:}Z) &= I(Y_1^*{:}Z) + I(Y{:}Y_1^*) \\ I(Y{:}Y_1^*) &= I(Y_2^*{:}Y_1^*) + I(Y{:}Y_2^*) \\ I(Y{:}Y_2^*) &= I(Y_3^*{:}Y_2^*) + I(Y{:}Y_3^*) \\ \cdots \quad & \quad \cdots \qquad \cdots \\ I(Y{:}Y_{n-1}^*) &= I(Y_n^*{:}Y_{n-1}^*) + I(Y{:}Y_n^*) \end{aligned} \tag{5.21}$$

where $Y \in \mathbf{Y}$, $Y_1^* \in \mathbf{Y}_1, \ldots, Y_n^* \in \mathbf{Y}_n$, and

$$\begin{aligned} Y_1^*(\omega) &= \frac{exp\,(t_1 T_1(\omega)}{M_1(t_1)} Z(\omega), M_1(t_1) = \int exp(t_1 T_1(\omega))\, Z(\omega)\, dP(\omega) \\ &= M(t_1), \\ \theta_1 &= \frac{\partial}{\partial t_1} \ell n\, M_1(t_1), \end{aligned} \tag{5.22}$$

$$Y_2^*(\omega) = \frac{exp\,(t_2T_2(\omega))}{M_2(t_2)}\; Y_1^*(\omega) = \frac{exp\,(t_1T_1(\omega) + t_2T_2(\omega))\,Z(\omega)}{M_2(t_2)\,M_1(t_1)} \tag{5.23}$$

$$M_2(t_2) = \int exp\,(t_2T_2(\omega))\,Y_1^*(\omega)\,dP(\omega) = \int \frac{exp\,(t_1T_1(\omega) + t_2T_2(\omega))\,Z(\omega)\,dP(\omega)}{M_1(t_1)}$$

or $M_2(t_2)M_1(t_1) = \int exp\,(t_1T_1(\omega) + t_2T_2(\omega))Z(\omega)\,dP(\omega) = M(t_1, t_2)$,

$$\theta_2 = \frac{\partial}{\partial t_2}\,\ell n\, M_2(t_2) = \frac{\partial}{\partial t_2}\,\ell n\, M(t_1, t_2),$$

$$Y_3^*(\omega) = \frac{exp\,(t_3T_3(\omega))\,Y_2^*(\omega)}{M_3(t_3)} = \frac{exp\,(t_2T_2 + t_3T_3)\,Y_1^*}{M_3(t_3)\,M_2(t_2)}$$

$$= \frac{exp\,(t_1T_1 + t_2T_2 + t_3T_3)\,Z(\omega)}{M_3(t_3)\,M_2(t_2)\,M_1(t_1)},$$

$$M_3(t_3) = \int exp(t_3T_3)\,Y_2^*\,dP \tag{5.24}$$

$$= \int \frac{exp\,(t_2T_2 + t_3T_3)\,Y_1^*\,dP}{M_2(t_2)} = \int \frac{exp\,(t_1T_1 + t_2T_2 + t_3T_3)\,ZdP}{M_2(t_2)\,M_1(t_1)}$$

or $M_1(t_1)\,M_2(t_2)\,M_3(t_3) = \int exp\,(t_1T_1 + t_2T_2 + t_3T_3)\,ZdP = M(t_1, t_2, t_3)$

$$\theta_3 = \frac{\partial}{\partial t_3}\,\ell n\, M_3(t_3) = \frac{\partial}{\partial t_3}\,\ell n\, M(t_1, t_2, t_3), \textit{ etc.}$$

Note that

$$I(Y_r^*:Y_{r-1}^*) = t_r\theta_r - \ell n\, M_r(t_r) \tag{5.25}$$

and

$$I(Y^*:Z) = I(Y_1^*:Z) + I(Y_2^*:Y_1^*) + \ldots + I(Y_n^*:Y_{n-1}^*), \tag{5.26}$$

$$I(Y_r^*:Z) = I(Y_1^*:Z) + I(Y_2^*:Y_1^*) + \ldots + I(Y_r^*:Y_{r-1}^*) .$$

It follows from (5.21) that

$$I(Y:Y_1^*) \geq I(Y:Y_2^*) \geq \ldots \geq I(Y:Y_n^*) . \tag{5.27}$$

If we denote by β_r the sub-sigma-algebra induced by $T_1(\omega), T_2(\omega), \ldots, T_r(\omega)$, then (5.21) is equivalent to

$$\begin{aligned} I(\mathbf{A}) &= I(\beta_1) + I(\mathbf{A}|\beta_1) \\ I(\mathbf{A}|\beta_1) &= I(\beta_2|\beta_1) + I(\mathbf{A}|\beta_2) \\ &\vdots \\ I(\mathbf{A}|\beta_{n-1}) &= I(\beta_n|\beta_{n-1}) + I(\mathbf{A}|\beta_n) \end{aligned} \tag{5.28}$$

and (5.26) is equivalent to

$$I(\beta_n) = I(\beta_1) + I(\beta_2|\beta_1) + \ldots + I(\beta_n|\beta_{n-1}) . \tag{5.29}$$

For applications of the foregoing concepts to frequency or count data dealing with categorical variables using the principle of minimum discrimination information estimation, and the analysis of information on nested models see Gokhale and Kullback (1978), Kullback and Keegel (1984).

III: LIMITING PROPERTIES

We shall consider sequences of discrimination information values and examine limiting properties depending on the behavior of sequences of generalized probability densities, sequences of probability measures, and sequences of sub - sigma - algebras. As before we consider the probability space $(\Omega, \mathbf{A}, P)$, and use the notation of section 3 of Chapter I.

1. Lower Semi-continuity

For the present we shall only assume that

$$\lim_{n\to\infty} \mu_n (A) = \mu (A), \quad \lim_{n\to\infty} \mu_{tn} (A) = \mu_t (A), \quad A \in \mathbf{A} \tag{1.1}$$

and we do not require that the convergence be uniform in $A \in \mathbf{A}$.
Let us write

$$\begin{aligned} I(\mu_{tn} : \mu_n) &= \int Z_{tn} \, ln \frac{Z_{tn}}{Z_n} dP = \int d\mu_{tn} \, ln \frac{d\mu_{tn}}{d\mu_n} \\ &= \int W_{tn} \, ln \, W_{tn} \, d\mu_n = \int (log \, W_{tn}) \, d\mu_{tn}, \end{aligned} \tag{1.2}$$

$$\begin{aligned} I(\mu_t : \mu) &= \int Z_t \, ln \frac{Z_t}{Z} dP = \int d\mu_t \, ln \frac{d\mu_t}{d\mu} = \int W_t \, ln \, W_t \, d\mu \\ &= \int (ln \, W_t) \, d\mu_t . \end{aligned} \tag{1.3}$$

Theorem 1.1 If the probability measures μ_{tn}, μ_n satisfy (1.1) then

$$\lim_{n\to\infty} \inf I(\mu_{tn} : \mu_n) \geq I(\mu_t : \mu), \tag{1.4}$$

that is, the functional I is lower semi-continuous, in other words if $C \geq I(\mu_{tn} : \mu_n)$, then $C \geq I(\mu_t : \mu)$. (See Dobrushin (1960), Gelfand et al (1956), Kullback (1959, pp. 70-71).)

Proof. Using the convexity property,

$$\int Z_{tn} \ln \frac{Z_{tn}}{Z_n} dP \geq \sum_i \mu_{tn}(A_i) \ln \frac{\mu_{tn}(A_i)}{\mu_n(A_i)} \tag{1.5}$$

where the sum is taken over any finite **A** - measurable partition of Ω. Since the right-hand side of (1.5) is a continuous function of the μ's

$$\lim_{n\to\infty} \inf \int Z_{tn} \ln \frac{Z_{tn}}{Z_n} dP \geq \sum_i \mu_t(A_i) \ln \frac{\mu_t(A_i)}{\mu(A_i)}. \tag{1.6}$$

But as was indicated in the integral representation theorem 1.1 of Chapter II

$$\int Z_t \ln \frac{Z_t}{Z} dP = \operatorname{Sup} \sum_i \mu_t(A_i) \ln \frac{\mu_t(A_i)}{\mu(A_i)} \tag{1.7}$$

where the sup ($l.u.b.$) is taken over all possible **A**- measurable finite partitions of Ω. Hence it is seen from (1.6) that

$$\lim_{n\to\infty} \inf \int Z_{tn} \ln \frac{Z_{tn}}{Z_n} dP \geq \int Z_t \ln \frac{Z_t}{Z} dP. \tag{1.8}$$

Note from (1.8) that if $I(\mu_t : \mu) = \infty$, then $\lim_{n\to\infty} I(\mu_{tn} : \mu_n) = \infty$. Note also that if we only consider the relations in (3.4) of chapter I and use (1.1), that

if $\lim_{n\to\infty}\inf I(\mu_{tn} : \mu_n) < \infty$, then also $I(\mu_t : \mu) < \infty$ which implies that $\mu_t \ll \mu$.

Corollary 1.1. If the conditions of theorem 1.1 are satisfied and there exists a sufficient partition for all $n = 1,2,\ldots$, then

$$\lim_{n\to\infty} I(\mu_{tn} : \mu_n) = I(\mu_t : \mu). \tag{1.9}$$

Proof. If the partition in (1.5) is sufficient, there is equality, hence

$$\lim_{n\to\infty} \int Z_{tn} \, ln \frac{Z_{tn}}{Z_n} \, dP = \lim_{n\to\infty} \sum_i \mu_{tn}(A_i) \, ln \frac{\mu_{tn}(A_i)}{\mu_n(A_i)}$$

$$= \sum_i \mu_t(A_i) \, ln \frac{\mu_t(A_i)}{\mu(A_i)} \leq \int Z_t \, ln \frac{Z_t}{Z} \, dP. \tag{1.10}$$

The result in (1.10) combined with (1.4) then yields the desired (1.9).

We shall need two results from probability theory (see for example, Loève (1955), p.140, prob. 16, 17) as we begin to consider the limiting behavior depending on the Z's.

Lemma 1.1. $\int |Z_{tn} - Z_t| \, dP \to 0$, *resp.* $\int |Z_n - Z| \, dP \to 0$ as $n \to \infty$, if and only if $\int_A Z_{tn} \, dP \to \int_A Z_t \, dP$, *resp.* $\int_A Z_n \, dP \to \int_A Z \, dP$ as $n \to \infty$ uniformly in $A \in \mathbf{A}$

Lemma 1.2 If $Ztn \to Zt$ *resp.* $Zn \to Z$ in probability, then $\int_A Z_t \, dP \to \int_A Z_t \, dP$ *resp.* $\int_A Z_n \, dP \to \int_A Z dP$ as $n \to \infty$ uniformly in $A \in \mathbf{A}$. The convergence in probability may be replaced by almost sure convergence.

Note that Lemmas 1.1 and 1.2 provide the chain of implications

$$\mu_n(A) \to \mu(A), \text{ uniformly in } A \in \mathbf{A} \Leftrightarrow Z_n \to Z \text{ in } L_1 \text{ convergence} \Rightarrow \tag{1.11}$$

$Z_n \to Z$ in probability, $\Rightarrow \mu_n(A) \to \mu(A)$, uniformly in $A \in \mathbf{A}$ and a similar one with the subscript t, where

$$Z_n \to Z \text{ in } L_1 \text{ convergence } \Leftrightarrow \int |Z_n - Z| \, dP \to 0. \tag{1.12}$$

It is thus seen that even if the convergence in (1.1) is required to be uniform in $A \in \mathbf{A}$ with the consequent implication in (1.11) about the convergence of the Z's, we still have the result in Theorem 1.1 and need stronger conditions to get convergence to a limit.

We note here that in Dobrushin (1960) it is shown that (1.9) holds under the conditions (1.11) if and only if the sequence of the $ln\ W_{tn}$ is uniformly integrable. For other convergence criteria see Ikeda (1960), Kullback (1968).

2. Sequences of Sub-sigma-algebras

Suppose now that there is an increasing sequence of sub-sigma-algebras of $\mathbf{A}$ such that

$$\cdots \subset \beta_n \subset \beta_{n+1} \subset \cdots\cdot \subset \beta \subset \mathbf{A} \tag{2.1}$$

where β is the minimal sigma-algebra containing $\cup_n \beta_n$, usually denoted by $\beta_n \uparrow \beta$, or $\beta = \vee \beta_n$. We shall first prove

Theorem 2.1. $I(\beta_{n+1}) \geq I(\beta_n)$ with equality if and only if β_n is sufficient for β_{n+1}.

Proof:

$$I(\beta_{n+1}) = \int E^{\beta_{n+1}} Z_t \, ln \frac{E^{\beta_{n+1}} Z_t}{E^{\beta_{n+1}} Z} \, dP =$$

$$= \int E^{\beta_n} \left(E^{\beta_{n+1}} Z_t \ln \frac{E^{\beta_{n+1}} Z_t}{E^{\beta_{n+1}} Z} \right) dP \tag{2.2}$$

$$\geq \int E^{\beta_n} E^{\beta_{n+1}} Z_t \ln \frac{E^{\beta_n} E^{\beta_{n+1}} Z_t}{E^{\beta_n} E^{\beta_{n+1}} Z} dP = \int E^{\beta_n} Z_t \ln \frac{E^{\beta_n} Z_t}{E^{\beta_n} Z} dP$$

$$= I(\beta_n)$$

with equality if and only if

$$\frac{E^{\beta_{n+1}} Z_t}{E^{\beta_{n+1}} Z} = \frac{E^{\beta_n} Z_t}{E^{\beta_n} Z}. \tag{2.3}$$

The proof is essentially the same as that of Theorem 2.1 in Chapter II. We have also used the smoothing property of conditioning (Loève(1955, p. 351)). In Theorem 2.1 of Chapter II it was noted that for β_0 the sub-sigma-algebra generated by $(\emptyset, \Omega)$, $I(\beta_0) = 0$. Thus corresponding to the sequence of sub-sigma-algebras in (2.1), there is the following chain of inequalities for the discrimination information values

$$0 = I(\beta_o) \leq I(\beta_1) \leq \ldots \leq I(\beta_n) \leq I(\beta_{n+1}) \leq \ldots \leq I(\beta) \leq I(\mathbf{A}). \tag{2.4}$$

We shall now investigate conditions under which we can assert that

$$\lim_{n \to \infty} I(\beta_n) = I(\beta). \tag{2.5}$$

We shall first consider procedures which depend on the use of results from martingale theory and the proof of the integral representation theorem 1.1 of

Chapter II by using theorem 1.1 of this chapter. We shall then examine an approach which is more intrinsically information-theoretic in nature and derive (2.5) and the integral representation theorem without using results from martingale theory.

We shall need the following results from martingale theory. In Lemma 2.1 are stated conclusions which are consequences of Theorem 4.1, page 319 and Theorem 4.3, page 331, of Doob (1953).

Lemma 2.1. For the sequence of sub-sigma-algebras as in (2.1)

$$E^{\beta_n} Z_t \overset{a.s.}{\to} E^{\beta} Z_t , \; E^{\beta_n} Z \overset{a.s.}{\to} E^{\beta} Z , \; n \to \infty \tag{2.6}$$

$$\int | E^{\beta_n} Z_t - E^{\beta} Z_t | \, dP \to 0 , \; \int | E^{\beta_n} Z - E^{\beta} Z | \, dP \to 0 , \; n \to \infty . \tag{2.7}$$

In Neveu (1965), p. 51 it is shown that

$$\int | X_n - X | \, dP \to 0 , \; P (A_n \, \Delta \, A) \to 0 , \; n \to \infty \tag{2.8}$$

imply

$$\int_{A_n} X_n \, dP \to \int_A X dP , \; n \to \infty , \tag{2.9}$$

where $A_n \, \Delta \, A$ is the symmetric difference $A_n A^c + A_n^c A$. Since $B_n \in \beta_n , B \in \beta , P (B_n \, \Delta \, B) \to 0 , n \to \infty$ is a consequence of (2.1) (see, for example, Chung (1968, p. 232), Dobrushin (1959), Kallianpur (1960), Pérez (1957)) we have from (2.7)

$$\mu(B_n) = \int_{B_n} Z\,dP = \int_{B_n} E^{\beta_n} Z\,dP \to \int_B E^{\beta} Z\,dP$$

$$= \int_B Z\,dP = \mu(B), \; B_n \in \beta_n, B \in \beta. \tag{2.10}$$

We are now in a position to apply Theorem 1.1 with a slight change of notation. Let us set

$$Z_{tn} = E^{\beta_n} Z_t, \; Z_t = E^{\beta} Z_t, \; Z_n = E^{\beta_n} Z, \; Z = E^{\beta} Z \tag{2.11}$$

and treat β as the $\mathbf{A}$ in Theorem 1.1. Since

$$I(\mu_{tn} : \mu_n) = I(\beta_n), \; I(\mu_t : \mu) = I(\beta) \tag{2.12}$$

we can write (1.4) as

$$\lim_{n\to\infty} \inf I(\beta_n) \geq I(\beta). \tag{2.13}$$

Consequently if $I(\beta) = \infty$, $\lim\limits_{n\to\infty} I(\beta_n) = \infty$. If $I(\beta) < \infty$, we have from the non-decreasing sequence of values in (2.4) that $\lim\limits_{n\to\infty} I(\beta_n) \leq I(\beta)$, that is $\lim\limits_{n\to\infty} I(\beta_n) \leq I(\beta) \leq \lim\limits_{n\to\infty} \inf I(\beta_n)$, or the result in (2.5) in which we were interested.

We may also state

$$\lim_{n\to\infty} I(\beta_n) < \infty \Rightarrow I(\beta) < \infty \Rightarrow \mu_t \ll \mu. \tag{2.14}$$

Note that if for some n, β_n is sufficient for β, that is $I(\beta_n) = I(\beta)$, then for all $N \geq n, I(\beta_N) = I(\beta)$.

Let us now consider another approach (Hájek (1958)) which does use results from martingale theory but does not use a result depending on the integral representation theorem 1.1 of Chapter II and does lead to a proof of the integral representation theorem. We recall the notation in section 3 of Chapter I and section 2 of Chapter II as well as the sequence (2.1). We shall also need the result equivalent to (2.6) in the form

$$E_Z^{\beta_n} W_t \to E_Z^{\beta} W_t \quad a.s. \; n \to \infty . \tag{2.15}$$

We note that $\phi(W_t) = W_t \ln W_t$ is a function bounded below, since the value for which the convex function $\phi(W_t)$ is a minimum, as obtained from

$$\frac{d\phi}{dW_t} = 0 = \ln W_t + 1$$

occurs for $W_t = 1/e$ and $W_t \ln W_t \geq -1/e$. It also follows then that $E_Z^{\beta_n} W_t \ln E_Z^{\beta_n} W_t \geq -1/e$. The Fatou – Lebesgue Theorem then yields

$$\int E_Z^{\beta} W_t \ln E_Z^{\beta} W_t \, d\mu \leq \liminf_{n \to \infty} \int E_Z^{\beta_n} W_t \ln E_Z^{\beta_n} W_t \, d\mu . \tag{2.16}$$

But as in Theorem 2.1 Jensen's inequality yields

$$\int E_Z^{\beta} W_t \ln E_Z^{\beta} W_t \, d\mu \geq \int E_Z^{\beta_n} W_t \ln E_Z^{\beta_n} W_t \, d\mu \tag{2.17}$$

so that (2.16) and (2.17) imply

$$\lim_{n\to\infty} \int E_Z^{\beta_n} W_t \, ln \, E_Z^{\beta_n} W_t \, d\mu = \int E_Z^{\beta} W_t \, ln \, E_Z^{\beta} W_t \, d\mu \tag{2.18}$$

that is

$$\lim_{n\to\infty} I(\beta_n) = I(\beta). \tag{2.19}$$

Note that in deriving (2.19) no use was made of a result depending on the integral representation theorem 1.1 of Chapter II as was done in the previous derivation. We are then in a position to apply (2.19) to arrive at a proof of the integral representation theorem.

In section 2 of Chapter I, it was seen that a separable sigma-algebra can be generated by a regular sequence of partitions, and using the notation of (2.4) of Chapter I and (2.1) we have

$$\underline{\varepsilon}^n \uparrow \underline{\varepsilon} \tag{2.20}$$

According, as a special case of (2.5) we have that

$$\lim_{n\to\infty} I(\underline{\varepsilon}^n) = I(\underline{\varepsilon}). \tag{2.21}$$

Recalling (1.11), (2.15) and Theorem 3.1, all of Chapter II we note that (2.21) is the integral representation theorem, that is

$$I(\mathbf{A}) = \text{Sup} \sum \mu_t(A_i) \, ln \, \frac{\mu_t(A_i)}{\mu(A_i)} \tag{2.22}$$

where the sup is taken over all possible $\mathbf{A}$ -measurable finite partitions of Ω, and in particular if the β_n are generated by finite partitions then (2.19) is the same as (2.4) of Chapter II.

3. An Information-theoretic Approach

We shall now consider an approach to the results of section 2 which is believed to be of interest in that it is intrinsically information-theoretic in nature and is intended as a novel and structurally intuitive motivation of the integral representation theorem.

Suppose we have the sequence of sub-sigma-algebras

$$\beta_o \subset \beta_1 \subset \beta_2 \subset \ldots \subset \beta_n \uparrow \beta.$$

Consider

$$\int E^{\beta} Z \ln \frac{E^{\beta} Z}{E^{\beta_n} Z}\, dP = \int E^{\beta} Z \ln \frac{E^{\beta_{n+1}} Z}{E^{\beta_n} Z}\, dP + \int E^{\beta} Z \ln \frac{E^{\beta} Z}{E^{\beta_{n+1}} Z}\, dP \tag{3.1}$$

so that since

$$\int E^{\beta} Z \ln \frac{E^{\beta_{n+1}} Z}{E^{\beta_n} Z}\, dP = \int E^{\beta_{n+1}} \left(E^{\beta} Z \ln \frac{E^{\beta_{n+1}} Z}{E^{\beta_n} Z} \right) dP$$

$$= \int E^{\beta_{n+1}} Z \ln \frac{E^{\beta_{n+1}} Z}{E^{\beta_n} Z}\, dP \geq 0\,, \tag{3.2}$$

(recall that $E^{\beta_{n+1}}(E^{\beta} Z) = E^{\beta_{n+1}} Z$ and $E^{\beta_{n+1}}(E^{\beta_n} Z) = (E^{\beta_n} Z)$). We may write (3.1) as

$$\int E^{\beta} Z \ln \frac{E^{\beta} Z}{E^{\beta_n} Z} dP = \int E^{\beta_{n+1}} Z \ln \frac{E^{\beta_{n+1}} Z}{E^{\beta_n} Z} dP +$$

$$\int E^{\beta} Z \ln \frac{E^{\beta} Z}{E^{\beta_{n+1}} Z} dP . \qquad (3.3)$$

Hence using the notation

$$I(\beta : \beta_n) = \int E^{\beta} Z \ln \frac{E^{\beta} Z}{E^{\beta_n} Z} dP \qquad (3.4)$$

we have the chain of inequalities

$$I(\beta : \beta_o) \geq I(\beta : \beta_1) \geq \ldots \geq I(\beta : \beta_n) \geq I(\beta : \beta_{n+1}) \geq \ldots \geq 0 \qquad (3.5)$$

where it is seen from (3.3) that $I(\beta : \beta_n) = I(\beta : \beta_{n+1}) \Leftrightarrow I(\beta_{n+1} : \beta_n) = 0$. Since the monotonically decreasing sequence (3.5) is bounded below it converges and hence

$$I(\beta : \beta_n) - I(\beta : \beta_{n+1}) = \int E^{\beta_{n+1}} Z \ln \frac{E^{\beta_{n+1}} Z}{E^{\beta_n} Z} dP \to 0, n \to \infty . \qquad (3.6)$$

Hence using the result in Kullback (1967), (3.6) implies

$$\int | E^{\beta_{n+1}} Z - E^{\beta_n} Z | \, dP \to 0, \quad n \to \infty . \qquad (3.7)$$

Indeed, by considering $I(\beta : \beta_{n+m}) - I(\beta : \beta_n)$ we can also get

$$\int | E^{\beta_{n+m}} Z - E^{\beta_n} Z | \, dP \to 0, \; m, n \to \infty . \tag{3.8}$$

Hence the sequence $E^{\beta_n} Z$ is L_1 fundamental and there exists an $X \in L_1$ such that

$$\int | E^{\beta_n} Z - X | \, dP \to 0, \quad n \to \infty \tag{3.9}$$

and

$$\int E^{\beta_n} Z \, dP \to \int X dP \, , \quad n \to \infty \, , \tag{3.10}$$

that is

$$\int X dP = 1 \tag{3.11}$$

since $\int E^{\beta_n} Z dP = 1$ for all n (see Loève (1955), pp. 157, 161).

Without restricting the generality we may take X to be β - measurable by the following argument. Using the result in Loève (1955, p. 348) that

$$X_n \to X \text{ in } L_1 \text{ convergence} \Leftrightarrow E^{\beta} X_n \to E^{\beta} X \text{ in } L_1 \text{ convergence}$$

we have

$$\int | E^{\beta_n} Z - X | \, dP \to 0 \, , \; n \to \infty \Rightarrow \int | E^{\beta} E^{\beta_n} Z - E^{\beta} X | \, dP \to 0 \;\; n \to \infty$$

$$\Rightarrow \int | E^{\beta_n} Z - E^{\beta} X | \, dP \to 0, \; n \to \infty$$

since $E^{\beta} E^{\beta_n} Z = E^{\beta_n} Z$. If X is β – measurable $E^{\beta} X = X$ a.s.

We shall now show that $X = E^{\beta} Z$ a.s. Applying lemma 1.1 to (3.9) we have

$$\int_A E^{\beta_n} Z\, dP \to \int_A X dP. \tag{3.12}$$

If $A \in \beta_k$, then for $n \geq k$

$$\int_A E^{\beta_n} Z dP = \int_A E^{\beta} Z dP \tag{3.13}$$

so that

$$\int_A X dP = \int_A E^{\beta} Z dP . \tag{3.14}$$

Since (3.14) is true for $A \in \beta_k$, it is true for $A \in \cup \beta_k$. Thus the probability measures defined by the integrals in (3.14) are identical on the field $\cup \beta_k$ and hence by the extension theorem Loève (1955, p. 87) (3.14) holds for $A \in \beta$. The Radon-Nikodym theorem then yields from (3.14) that

$$X = E^{\beta} Z [P_{\beta}] \tag{3.15}$$

Since

$$\int | E^{\beta_n} Z - E^{\beta} Z | dP \leq \int | E^{\beta_n} Z - X | dP + \int | X - E^{\beta} Z | dP \tag{3.16}$$

we see from (3.9) and (3.15) that

$$\int | E^{\beta_n} Z - E^{\beta} Z | dP \to 0, \quad n \to \infty \tag{3.17}$$

that is (2.7) in lemma 2.1 .

$$\int E^{\beta} Z \, ln \, \frac{E^{\beta} Z}{X} \, dP = 0 \tag{3.18}$$

if and only if $X = E^{\beta} Z a.s.$, then in view of (3.15) we conjecture that the sequence (3.5) has the limit zero, that is,

$$\lim_{n \to \infty} I(\beta : \beta_n) = 0. \tag{3.19}$$

We show that (3.19) is true as follows. We write

$$I(\beta : \beta_n) = \int E^{\beta} Z \, ln \, \frac{E^{\beta} Z}{E^{\beta_n} Z} \, dP = \int E^{\beta} Z \, ln \, E^{\beta} Z \, dP - \int E^{\beta} Z \, ln \, E^{\beta_n} Z \, dP$$

$$= \int E^{\beta} Z \, ln \, E^{\beta} Z dP - \int E^{\beta_n} Z \, ln \, E^{\beta_n} Z dP .$$

We have shown that $E^{\beta_n} Z \to E^{\beta} Z$ in L_1 convergence which implies that $E^{\beta_n} Z \to E^{\beta} Z$ in probability. The convergence in probability implies that there exists a sequence $\{n_k\}$ of integers increasing to infinity such that (Loève (1955), p. 151)

$$E^{\beta_{n_k}} Z \to E^{\beta} Z \ a.s..$$

Since the convex function $E^{\beta_n} Z \ln E^{\beta_n} Z \geq -1/e$ the Fatou – Lebesgue Theorem (Loève (1955), p. 152) yields

$$\lim\inf \int E^{\beta_{n_k}} Z \ln E^{\beta_{n_k}} Z \, dP \geq \int E^{\beta} Z \ln E^{\beta} Z \, dP. \tag{3.20}$$

But the convexity, Jensen's inequality, and the smoothing property of conditioning (Loève (1955) , pp. 348, 351) lead to

$$\int E^{\beta} Z \ln E^{\beta} Z dP \geq \int E^{\beta_n} Z \ln E^{\beta_n} Z dP \geq \int E^{\beta_m} Z \ln E^{\beta_m} Z dP \tag{3.21}$$

for all n, m such that $n > m$.

From the monotonic property in (3.21) combined with (3.20) we conclude that

$$\lim \int E^{\beta_{n_k}} Z \ln E^{\beta_{n_k}} Z dP = \int E^{\beta} Z \ln E^{\beta} Z dP$$

or

$$\lim I\,(\beta : \beta_{n_k}) = 0.$$

Since $I(\beta : \beta_n)$ converges, it must converge to the same limit, that is, we have (3.19).

Note that if we interpret $I(\beta : \beta_n)$ as a measure of the closeness of the approximation to the measures over β by the measures over β_n the sequence (3.5) implies that the approximation gets better as n gets larger and in the limit the approximation is exact to within sets of measure zero.

A similar argument of course follows for Z_t. If we start with

$$\int E_Z^{\beta} W_t \ln \frac{E_Z^{\beta} W_t}{E_Z^{\beta_n} W_t} d\mu = \int E_Z^{\beta} W_t \ln \frac{E_Z^{\beta_{n+1}} W_t}{E_Z^{\beta_n} W_t} d\mu +$$

$$\int E_Z^{\beta} W_t \, ln \, \frac{E_Z^{\beta} W_t}{E_Z^{\beta_{n+1}} W_t} \, d\mu \tag{3.22}$$

since

$$\int E_Z^{\beta} W_t \, ln \, \frac{E_Z^{\beta_{n+1}} W_t}{E_Z^{\beta_n} W_t} \, d\mu = \int E_Z^{\beta_{n+1}} W_t \, ln \, \frac{E_Z^{\beta_{n+1}} W_t}{E_Z^{\beta_n} W_t} \, d\mu \tag{3.23}$$

we can repeat the preceding argument and conclude that

$$\int | E_Z^{\beta_n} W_t - Y_t | \, d\mu \to 0 \, , \; n \to \infty$$

$$\int Y_t \, d\mu = 1 \, ,$$

$$Y_t = E_Z^{\beta} W_t [\mu_\beta] \tag{3.24}$$

$$\lim_{n\to\infty} \int E_Z^{\beta} W_t \, ln \, \frac{E_Z^{\beta} W_t}{E_Z^{\beta_n} W_t} \, d\mu = 0 \, . \tag{3.25}$$

We shall now use these results to show that

$$I(\beta) = \int E_Z^{\beta} W_t \, ln \, E_Z^{\beta} W_t \, d\mu = \lim_{n\to\infty} \int E_Z^{\beta_n} W_t \, ln \, E_Z^{\beta_n} W_t \, d\mu = \lim_{n\to\infty} I(\beta_n) . \tag{3.26}$$

Using the result in theorem 2.3 of Chapter II we may write

$$I(\beta | \beta_o) = I(\beta_1 | \beta_o) + I(\beta | \beta_1)$$

$$I(\beta | \beta_1) = I(\beta_2 | \beta_1) + I(\beta | \beta_2)$$

$$\begin{aligned} &\cdots \quad \cdots \quad \cdots \\ &I(\beta \mid \beta_k) = I(\beta_{k+1} \mid \beta_k) + I(\beta \mid \beta_{k+1}) \\ &\cdots \quad \cdots \quad \cdots \end{aligned} \tag{3.27}$$

or using the relations in (3.27)

$$I(\beta \mid \beta_o) = I(\beta) = \sum_{k=o}^{n} I(\beta_{k+1} \mid \beta_k) + I(\beta \mid \beta_{n+1}) \tag{3.28}$$

where

$$I(\beta \mid \beta_N) = \int E_Z^{\beta} W_t \, ln \frac{E_Z^{\beta} W_t}{E_Z^{\beta_N} W_t} \, d\mu. \tag{3.29}$$

Since the relation (3.25) applied to (3.29) implies

$$I(\beta \mid \beta_N) \to 0 \text{ as } N \to \infty \tag{3.30}$$

we get from (3.28) (see (2.8), (2.10) of Chapter II)

$$I(\beta) = \sum_{k=o}^{\infty} I(\beta_{k+1} \mid \beta_k) = \sum_{k=o}^{\infty} (I(\beta_{k+1}) - I(\beta_k)) = \lim_{n\to\infty} I(\beta_n). \tag{3.31}$$

From (3.31) it is seen that if $I(\beta) < \infty$ then $I(\beta_{n+1} \mid \beta_n) \to 0$, $n \to \infty$.
If $I(\beta_{n+1} \mid \beta_n) \to I < \infty$ as $n \to \infty$, then by the Toeplitz lemma

$$\lim_{n\to\infty} \frac{1}{n} I(\beta_n) = I.$$

Note that in this section we have had no need to use any result depending on the integral representation theorem 1.1 of Chapter II. Thus the same argument as that following (2.19) at the end of the preceding section leads to the integral representation theorem. See Kullback and Thall (1977).

For application of the preceding results, particularly (3.31) or (2.19) to stochastic processes we state in Lemma 3.1 a result which is Theorem 1.6, page 604 of Doob (1953).

Lemma 3.1. Let **A** be a sigma-algebra of ω sets, and let $\{x(t,\omega), t \in T\}$ be a family of ω functions measurable with respect to **A**. Let β_S be the sigma-algebra generated by $\{x(t,\omega), t \in S \subset T\}$. Suppose that T is non-denumerable. Then if $\Lambda \in \beta_T$ there is a denumerable subset S (depending on Λ) of T, such that $\Lambda \in \beta_S$. If $X(\omega)$ is an ω function measurable with respect to β_T, there is a denumerable subset S (depending on X) such that X is measurable with respect to β_S.

Now let $\{x(t,\omega), t \in T\}$ be an arbitrary system of random variables defining a stochastic process. Let β_N be the sigma-algebra generated by the sub-system $\{x(t,\omega), t \in N \subset T\}$ and β_T the sigma-algebra generated by the system of random variables defining the stochastic process. We can now state

Theorem 3.1.

$$I(\beta_T) = \sup_{N \in \eta} I(\beta_N) \tag{3.32}$$

where η is the class of all finite subsets of T.

Proof. If T is countable, then it is possible to choose finite subsets $N_1 \subset N_2 \subset \ldots$ such that β_T is the smallest sigma-algebra containing $\bigcup_{i=1}^{\infty} \beta_{N_i}$ and (3.32) is then essentially (3.31) or (2.19). If T is not countable we shall use Lemma 3.1. If $I(\beta_S) = \infty$, then $I(\beta_T) = \infty$. Suppose $I(\beta_T) < \infty$, then $E_Z^{\beta_T} W_t = E^{\beta_T} Z_t / E^{\beta_T} Z$ (see (3.5) of Chapter I) exists and is of course measurable with respect to β_T. But according to Lemma 3.1 every function measurable with respect to β_T is measurable with respect to β_S for at least one countable $S \subset T$ so that

$$E_Z^{\beta_T} W_t = E_Z^{\beta_S} W_t[\mu] \,, \quad \text{or} \quad E^{\beta_T} Z_t \,/\, E^{\beta_T} Z = E^{\beta_S} Z_t \,/\, E^{\beta_S} Z \; a.s. \tag{3.33}$$

Since (3.33) is the necessary and sufficient condition that $I(\beta_S) = I(\beta_T)$ we get

$$I(\beta_T) = I(\beta_S) = \sup_{N \in \eta} I(\beta_N). \tag{3.34}$$

For a similar result see Hájek (1958) .

4. Monotonicity

If the sigma-algebra β in (2.1) is not the sigma-algebra $\mathbf{A}$ of the probability space $(\Omega, \mathbf{A}, P)$ then again using theorem 2.3 of Chapter II as in (3.27) we have

$$I(\mathbf{A}) = I(\beta) + I(\mathbf{A} \mid \beta) \tag{4.1}$$

$$I(\mathbf{A}) = I(\beta_n) + I(\mathbf{A} \mid \beta_n). \tag{4.2}$$

We can now derive certain limiting relations in which $\mathbf{A}$ plays a role. From (4.2) and (3.31) we see that

$$\begin{aligned} I(\mathbf{A}) &= \lim_{n\to\infty} I(\beta_n) + \lim_{n\to\infty} I(\mathbf{A} \mid \beta_n) \\ &= I(\beta) + \lim_{n\to\infty} I(\mathbf{A} \mid \beta_n), \end{aligned} \tag{4.3}$$

hence for $I(\mathbf{A}) < \infty$

$$I(\mathbf{A}) - I(\beta) = I(\mathbf{A} \mid \beta) = \lim_{n\to\infty} I(\mathbf{A} \mid \beta_n). \tag{4.4}$$

Similarly from

$$I(\beta) = I(\beta_n) + I(\beta \mid \beta_n) \tag{4.5}$$

$$I(\beta) = \lim_{n\to\infty} I(\beta_n) + \lim_{n\to\infty} I(\beta \mid \beta_n) \tag{4.6}$$

so that if $I(\beta) < \infty$

$$I(\beta) = \lim_{n\to\infty} I(\beta_n) \Leftrightarrow \lim_{n\to\infty} I(\beta \mid \beta_n) = 0. \tag{4.7}$$

Note that if $I(\mathbf{A}) < \infty$ and $\beta_m \subset \beta_n$, then using (4.2) we have

$$I(\mathbf{A} \mid \beta_m) - I(\mathbf{A} \mid \beta_n) = I(\beta_n) - I(\beta_m) \geq 0. \tag{4.8}$$

Similarly for $\beta_k \subset \beta_m \subset \beta_n$ we have

$$I(\beta_n \mid \beta_k) \geq I(\beta_m \mid \beta_k). \tag{4.9}$$

As a matter of fact for $\beta_1 \subset \beta_2 \subset \mathbf{A}$ we can write if $I(\mathbf{A}) < \infty$ (for a related discussion see Ghurye (1968))

$$I(\mathbf{A} \mid \beta_1) = I(\beta_2 \mid \beta_1) + I(\mathbf{A} \mid \beta_2) \tag{4.10}$$

which can be proven either directly or from the fact that

$$[\, I(\mathbf{A}) - I(\beta_1)\,] = [\, I(\beta_2) - I(\beta_1)\,] + [\, I(\mathbf{A}) - I(\beta_2)\,]. \tag{4.11}$$

Since the information values in (4.10) are nonnegative it follows that

$$I(\mathbf{A} \mid \beta_1) \geq I(\mathbf{A} \mid \beta_2) \tag{4.12}$$

with equality if and only if $I(\beta_2 \mid \beta_1) = 0$ and

$$I(\mathbf{A} \mid \beta_1) \geq I(\beta_2 \mid \beta_1) \tag{4.13}$$

with equality if and only if $\mathrm{I}(\mathbf{A} \mid \beta_2) = 0$ (cf (3.12) in Chapter II).

Using A in place of β in (3.27) we have corresponding to (3.28)

$$I(\mathbf{A} \mid \beta_o) = I(\mathbf{A}) = \sum_{k=o}^{n} I(\beta_{k+1} \mid \beta_k) + I(\mathbf{A} \mid \beta_{k+1}) \tag{4.14}$$

so that using (4.4)

$$I(\mathbf{A}) = \sum_{k=o}^{\infty} I(\beta_{k+1} \mid \beta_k) + I(\mathbf{A} \mid \beta). \tag{4.15}$$

IV: PRODUCT MEASURE SPACES

1. Generalized Contingency Tables

The following discussion will form the background for an application of certain of the results of statistical information theory. The results in this section for the particular case of discrete spaces are essentially those for contingency tables in Kullback (1959, Chapter 8). For a similar discussion see Dobrushin (1959), Pinsker (1964). We use the notation in Halmos (1950).

Consider two measure spaces $(\Omega_1, \mathbf{A}_1, \lambda_1)$, $(\Omega_2, \mathbf{A}_2, \lambda_2)$ and the product measure space $(\Omega_1 x \Omega_2, \mathbf{A}_1 x \mathbf{A}_2, \lambda_1 x \lambda_2)$ where λ_1 and λ_2 are sigma-finite, and $\lambda_1 x \lambda_2$ is also sigma-finite. Suppose $f(\omega_1, \omega_2)$ is a non-negative function, measurable on $(\Omega_1 x \Omega_2, \mathbf{A}_1 x \mathbf{A}_2)$ and such that

$$\mu(E) = \int_E f(\omega_1, \omega_2)\, d(\lambda_1 x \lambda_2 (\omega_1, \omega_2)),\ E \in \mathbf{A}_1 x \mathbf{A}_2 \tag{1.1}$$

is a probability measure on $(\Omega_1 x \Omega_2, \mathbf{A}_1 x \mathbf{A}_2)$. Let

$$g(\omega_1) = \int f(\omega_1, \omega_2)\, d\lambda_2(\omega_2),\quad h(\omega_2) = \int f(\omega_1, \omega_2)\, d\lambda_1(\omega_1) \tag{1.2}$$

or

$$g(\omega_1) = \int f_{\omega_1}(\omega_2)\, d\lambda_2(\omega_2),\quad h(\omega_2) = \int f^{\omega_2}(\omega_1)\, d\lambda_1(\omega_1) \tag{1.3}$$

where $f_{\omega_1}(\omega_2)$ and $f^{\omega_2}(\omega_1)$ are respectively the section of $f(\omega_1, \omega_2)$ by ω_1 and by ω_2 and $f_{\omega_1}(\omega_2)$ is defined on Ω_2 by $f_{\omega_1}(\omega_2) = f(\omega_1, \omega_2)$ and $f^{\omega_2}(\omega_1)$ is defined on Ω_1 by $f^{w_2}(\omega_1) = f(\omega_1, \omega_2)$ and $f_{\omega_1}(\omega_2)$ and $f^{\omega_2}(\omega_1)$ are respectively measurable $\mathbf{A}_2$, $\mathbf{A}_1$,. Since $f(\omega_1, \omega_2)$ is defined $[\lambda_1 \times \lambda_2]$, $g(\omega_1)$ is defined $[\lambda_1]$ and $h(\omega_2)$ is defined $[\lambda_2]$, where $[\cdot]$ means except for sets of measure zero with respect to the measure indicated as $\cdot$. If $E \in (\Omega_1 \, x \, \Omega_2, \mathbf{A}_1 \, x \, \mathbf{A}_2)$, then

$$
\begin{aligned}
E_{\omega_1} &= \{\omega_2 : (\omega_1, \omega_2) \in E\} \in (\Omega_2, \mathbf{A}_2) \\
E^{\omega_2} &= \{\omega_1 : (\omega_1, \omega_2) \in E\} \in (\Omega_1, \mathbf{A}_1).
\end{aligned}
\tag{1.4}
$$

In particular, if $E = A_1 \, x \, A_2$, then $E_{\omega_1} = A_2$ and $E^{\omega_2} = A_1$. For the indicator function of sets

$$
I_E(\omega_1, \omega_2) = I_{E_{\omega_1}}(\omega_2) = I_{E^{\omega_2}}(\omega_1) \tag{1.5}
$$

and if $E = A_1 \, x \, \Omega_2$, $I_E(\omega_1, \omega_2) = I_{A_1}(\omega_1)$, and if $E = \Omega_1 \, x \, A_2$, $I_E(\omega_1, \omega_2) = I_{A_2}(\omega_2)$.

The following results hold for the probability measure μ and associated probability measures,

$$
\begin{aligned}
\mu(A_1 \, x \, \Omega_2) &= \int_{A_1 \, x \, \Omega_2} f(\omega_1, \omega_2) \, d(\lambda_1 \, x \, \lambda_2(\omega_1, \omega_2)) \\
&= \int I_{A_1 \, x \, \Omega_2} f(\omega_1, \omega_2) \, d(\lambda_1 \, x \, \lambda_2(\omega_1, \omega_2)) \\
&= \int \{ \int I_{A_1 \, x \, \Omega_2} f(\omega_1, \omega_2) \, d\lambda_1(\omega_1) \} \, d\lambda_2(\omega_2) = \int \{ \int_{A_1} f(w_1, \omega_2) \, d\lambda_1(\omega_1) \} \, d\lambda_2(\omega_2) \\
&= \int_{A_1} \{ \int f((\omega_1, \omega_2) \, d\lambda_2(\omega_2) \} \, d\lambda_1(\omega_1) = \int_{A_1} g(\omega_1) \, d\lambda_1(\omega_1) = \mu_1(A_1),
\end{aligned}
\tag{1.6}
$$

and similarly

$$\mu(\Omega_1 x A_2) = \int_{\Omega_1 x A_2} f(\omega_1, \omega_2)\, d(\lambda_1 x \lambda_2((\omega_1, \omega_2))$$

$$= \int I_{\Omega_1 x A_2} f(\omega_1, \omega_2)\, d(\lambda_1 x \lambda_2(\omega_1, \omega_2)) \tag{1.7}$$

$$= \int_{A_2} \{ \int f(\omega_1, \omega_2)\, d\lambda_1(\omega_1) \}\, d\lambda_2(\omega_2)$$

$$= \int_{A_2} h(\omega_2)\, d\lambda_2(\omega_2) = \mu_2(A_2),$$

that is, μ_1 and μ_2 are the marginal probability measures of the probability measure μ. The probability product measure $\mu_1 \, x \, \mu_2$ can now be defined on $(\Omega_1 \, x \, \Omega_2, \mathbf{A}_1 \, x \, \mathbf{A}_2)$ and it follows that

$$\mu_1 x \mu_2(E) = \int_E g(\omega_1)\, h(\omega_2)\, d(\lambda_1 x \lambda_2(\omega_1, \omega_2)),\ E \in (\Omega_1 x \Omega_2, \mathbf{A}_1 x \mathbf{A}_2). \tag{1.8}$$

We note that $\mu \ll \lambda_1 \, x \, \lambda_2$, $\mu_1 \ll \lambda_1$, $\mu_2 \ll \lambda_2$, $\mu_1 \, x \, \mu_2 \ll \lambda_1 \, x \, \lambda_2$ and $\mu \ll \mu_1 \, x \, \mu_2$. Generally,

$$\mu(E) = \int_E f(\omega_1, \omega_2)\, d(\lambda_1 x \lambda_2(\omega_1, \omega_2))$$

$$= \int \{ \int I_E(\omega_1, \omega_2) f(\omega_1, \omega_2)\, d\lambda_2(\omega_2) \}\, d\lambda_1(\omega_1)$$

$$= \int \{ \int_{E_{\omega_1}} f(\omega_1, \omega_2)\, d\lambda_2(\omega_2) \}\, d\lambda_1(\omega_1) \tag{1.9}$$

$$= \int \{ \int_{E^{\omega_2}} f(\omega_1, \omega_2)\, d\lambda_1(\omega_1) \}\, d\lambda_2(\omega_2),$$

and

$$\mu_1 x \mu_2(E) = \int \{ \int_{E_{\omega_1}} h(\omega_2)\, d\lambda_2(\omega_2) \} g(\omega_1)\, d\lambda_1(\omega_1)$$

$$= \int \{ \int_{E^{\omega_2}} g(\omega_1)\, d\lambda_1(\omega_1) \} h(\omega_2)\, d\lambda_2(\omega_2) \tag{1.10}$$

$$= \int \mu_2(E_{\omega_1}) g(\omega_1)\, d\lambda_1(\omega_1) = \int \mu_1(E^{\omega_2}) h(\omega_2)\, d\lambda_2(\omega_2)$$

$$= \int \mu_2(E_{\omega_1})\, d\mu_1(\omega_1) = \int \mu_1(E^{\omega_2})\, d\mu_2(\omega_2).$$

Let us define

$$f(\omega_2 | \omega_1) = f(\omega_1, \omega_2) / g(\omega_1),\ f(\omega_1 | \omega_2) = f(\omega_1, \omega_2) / h(\omega_2) \tag{1.11}$$

$$\int_{E_{\omega_1}} f(\omega_2 | \omega_1)\, d\lambda_2(\omega_2) = \mu_2(E_{\omega_1} | \omega_1),\ \int_{E^{\omega_2}} f(\omega_1 | \omega_2)\, d\lambda_1(\omega_1) = \mu_1(E^{\omega_2} | \omega_2)$$

then from (1.9)

$$\mu(E) = \int \mu_2(E_{\omega_1} | \omega_1) g(\omega_1)\, d\lambda_1(\omega_1) = \int \mu_1(E^{\omega_2} | \omega_2) h(\omega_2)\, d\lambda_2(\omega_2)$$

$$= \int \mu_2(E_{\omega_1} | \omega_1)\, d\mu_1(\omega_1) = \int \mu_1(E^{\omega_2} | \omega_2)\, d\mu_2(\omega_2) \tag{1.12}$$

$$= \int_E f(\omega_2 | \omega_1) g(\omega_1)\, d(\lambda_1 x \lambda_2(\omega_1, \omega_2))$$

$$= \int_E f(\omega_1 | \omega_2) h(\omega_2)\, d(\lambda_1 x \lambda_2(\omega_1, \omega_2)).$$

Note that

$$\mu_2(E_{\omega_1} | \omega_1) = \int_{E_{\omega_1}} f(\omega_2 | \omega_1)\, d\lambda_2(\omega_2) = \int I_E(\omega_1, \omega_2) \frac{f(\omega_1, \omega_2)}{g(\omega_1)} d\lambda_2(\omega_2)$$

$$= \int I_E(\omega_1, \omega_2) \frac{f(\omega_1, \omega_2)}{g(\omega_1)\, h(\omega_2)} d\mu_2(\omega_2)\ [\mu_1(\omega_1)] \tag{1.13}$$

is the conditional probability of E given ω_1. (See Halmos (1950, p. 210(2), Neveu (1965), Chapter III.)

Hence, if we write $(\Omega_1 \times \Omega_2, \mathbf{A}_1 \times \mathbf{A}_2, \lambda_1 \times \lambda_2)$ as $(\Omega, \mathbf{A}, \lambda)$ and there are two different probability measures on $(\Omega, \mathbf{A})$ defined by

$$P_1(E) = \int_E Z_1(\omega)\, d\lambda(\omega) = \int_E f_1(\omega_1, \omega_2)\, d(\lambda_1 \times \lambda_2(\omega_1, \omega_2)) \tag{1.14}$$

$$P_2(E) = \int_E Z_2(\omega)\, d\lambda(\omega) = \int_E f_2(\omega_1, \omega_2)\, d(\lambda_1 \times \lambda_2(\omega_1, \omega_2))$$

for $E \in \mathbf{A} = \mathbf{A}_1 \times \mathbf{A}_2$, then (see theorem 2.3 of Chapter II)

$$I(\mathbf{A}_1 \times \mathbf{A}_2 : P_1 ; P_2) = \int dP_1 \ln \frac{dP_1}{dP_2} = \int Z_1(\omega) \ln \frac{Z_1(\omega)}{Z_2(\omega)} d\lambda(\omega)$$

$$= \int f_1(\omega_1, \omega_2) \ln \frac{f_1(\omega_1, \omega_2)}{f_2(\omega_1, \omega_2)} d(\lambda_1 \times \lambda_2(\omega_1, \omega_2)) \tag{1.15}$$

$$= \int g_1(\omega_1) \ln \frac{g_1(\omega_1)}{g_2(\omega_1)} d\lambda_1(\omega_1) + \int f_1(\omega_1, \omega_2) \ln \frac{f_1(\omega_2 | \omega_1)}{f_2(\omega_2 | \omega_1)} d(\lambda_1 \times \lambda_2(\omega_1, \omega_2))$$

$$= \int g_1(\omega_1) \ln \frac{g_1(\omega_1)}{g_2(\omega_1)} d\lambda_1(\omega_1) +$$

$$+ \int g_1(\omega_1) \{ \int f_1(\omega_2 | \omega_1) \ln \frac{f_1(\omega_2 | \omega_1)}{f_2(\omega_2 | \omega_1)} d\lambda_2(\omega_2) \} d\lambda_1(\omega_1)$$

$$= I(\mathbf{A}_1 : P_1 ; P_2) + I(\mathbf{A}_1 \times \mathbf{A}_2 | \mathbf{A}_1 : P_1 ; P_2) .$$

Now consider three measure spaces $(\Omega_1, \mathbf{A}_1, \lambda_1)$, $(\Omega_2, \mathbf{A}_2, \lambda_2)$, $(\Omega_3, \mathbf{A}_3, \lambda_3)$ and the product measure space $(\Omega_1 \, x \, \Omega_2 \, x \, \Omega_3, \mathbf{A}_1 \, x \, \mathbf{A}_2 \, x \, \mathbf{A}_3, \lambda_1 \, x \, \lambda_2 \, x \, \lambda_3)$ where λ_1, λ_2, λ_3 and $\lambda_1 \, x \, \lambda_2 \, x \, \lambda_3$ are sigma-finite. Let $f(\omega_1, \omega_2, \omega_3)$ be a non-negative function measurable on $(\Omega_1 \, x \, \Omega_2 \, x \, \Omega_3, \mathbf{A}_1 \, x \, \mathbf{A}_2 \, x \, \mathbf{A}_3)$ and such that

$$\mu_{123}(E) = \int_E f(\omega_1, \omega_2, \omega_3) \, d(\lambda_1 \, x \, \lambda_2 \, x \, \lambda_3(\omega_1, \omega_2, \omega_3)), E \in \mathbf{A}_1 \, x \, \mathbf{A}_2 \, x \, \mathbf{A}_3 \qquad (1.16)$$

is a probability measure on $(\Omega_1 \, x \, \Omega_2 \, x \, \Omega_3, \mathbf{A}_1 \, x \, \mathbf{A}_2 \, x \, \mathbf{A}_3)$. Let

$$\begin{aligned} g(\omega_1) &= \int f(\omega_1, \omega_2, \omega_3) \, d(\lambda_2 \, x \, \lambda_3(\omega_2, \omega_3)) = f(\omega_1 \, . \, .), \\ h(\omega_2) &= \int f(\omega_1, \omega_2, \omega_3) \, d(\lambda_1 \, x \, \lambda_3(\omega_1, \omega_3)) = f(. \, \omega_2 \, .), \qquad (1.17) \\ i(\omega_3) &= \int f(\omega_1, \omega_2, \omega_3) \, d(\lambda_1 \, x \, \lambda_2(\omega_1, \omega_2)) = f(. \, . \, \omega_3). \end{aligned}$$

Hence (using a notation resembling one standard for contingency tables)

$$\begin{aligned} \mu_{123}(A_1 \, x \, \Omega_2 \, x \, \Omega_3) &= \int_{A_1 \, x \, \Omega_2 \, x \, \Omega_3} f(\omega_1, \omega_2, \omega_3) \, d(\lambda_1 \, x \, \lambda_2 \, x \, \lambda_3(\omega_1, \omega_2, \omega_3)) \\ &= \int_{A_1} \left\{ \int f(\omega_1, \omega_2, \omega_3) \, d(\lambda_2 \, x \, \lambda_3(\omega_2, \omega_3)) \right\} d\lambda_1(\omega_1) \qquad (1.18) \\ &= \int_{A_1} g(\omega_1) \, d\lambda_1(\omega_1) = \mu_{1..}(A_1), A_1 \in \mathbf{A}_1 \end{aligned}$$

and in a similar manner

$$\mu_{123}(\Omega_1 \, x \, A_2 \, x \, \Omega_3) = \int_{A_2} h(\omega_2) \, d\lambda_2(\omega_2) = \mu_{.2.}(A_2), A_2 \in \mathbf{A}_2, \qquad (1.19)$$

$$\mu_{123}(\Omega_1 \, x \, \Omega_2 \, x \, A_3) = \int_{A_3} i(\omega_3) \, d\lambda_3(\omega_3) = \mu_{..3}(A_3), A_3 \in \mathbf{A}_3. \qquad (1.20)$$

Similarly, let

$$
\begin{aligned}
f(\omega_1, \omega_2, \cdot\ ) &= \int f(\omega_1, \omega_2, \omega_3)\, d\lambda_3(\omega_3) \\
f(\cdot\ , \omega_2, \omega_3) &= \int f(\omega_1, \omega_2, \omega_3)\, d\lambda_1(\omega_1) \\
f(\omega_1, \cdot\ , \omega_3) &= \int f(\omega_1, \omega_2, \omega_3)\, d\lambda_2(\omega_2)
\end{aligned}
\tag{1.21}
$$

then

$$
\begin{aligned}
\mu_{1\,2\,3}(E\,x\,\Omega_3) &= \int_{E\,x\,\Omega_3} f(\omega_1, \omega_2, \omega_3)\, d\,(\lambda_1\,x\,\lambda_2\,x\,\lambda_3(\omega_1, \omega_2, \omega_3)) \\
&= \int_E f(\omega_1, \omega_2, \cdot\ )\, d\,(\lambda_1\,x\,\lambda_2(\omega_1, \omega_2)) = \mu_{1\,2\,.}(E), E \in \mathbf{A}_1\,x\,\mathbf{A}_2 \\
\mu_{1\,2\,3}(\Omega_1\,x\,F) &= \int_{\Omega_1\,x\,F} f(\omega_1, \omega_2, \omega_3)\, d\,(\lambda_1\,x\,\lambda_2\,x\,\lambda_3(\omega_1, \omega_2, \omega_3)) \\
&= \int_F f(\cdot\ , \omega_2, \omega_3)\, d\,(\lambda_2\,x\,\lambda_3(\omega_2, \omega_3)) = \mu_{.\,2\,3}(F), F \in \mathbf{A}_2\,x\,\mathbf{A}_3 \\
\mu_{1\,2\,3}(\Omega_2\,x\,G) &= \mu_{1\,.\,3}(G), G \in \mathbf{A}_1\,x\,\mathbf{A}_3\,.
\end{aligned}
\tag{1.22}
$$

Also

$$
\mu_{1\,.\,.}\,x\mu_{.\,2\,.}\,x\mu_{.\,.\,3}(E) = \int_E g(\omega_1)\, h(\omega_2)\, i(\omega_3)\, d\,(\lambda_1\,x\,\lambda_2\,x\,\lambda_3(\omega_1, \omega_2, \omega_3)),
\quad E \in \mathbf{A}_1\,x\,\mathbf{A}_2\,x\,\mathbf{A}_3,
\tag{1.23}
$$

$$
\mu_{1\,.\,.}\,x\,\mu_{.\,2\,3}(E) = \int_E g(\omega_1)\, f(\cdot\ , \omega_2, \omega_3)\, d\,(\lambda_1\,x\,\lambda_2\,x\,\lambda_3(\omega_1, \omega_2, \omega_3)),
\quad E \in \mathbf{A}_1\,x\,\mathbf{A}_2\,x\,\mathbf{A}_3,
\tag{1.24}
$$

etc.

Now

$$\mu_{1\,2\,3}(E) = \int_E f(\omega_1, \omega_2, \omega_3)\, d\,(\lambda_1 x \lambda_2 x \lambda_3 (\omega_1, \omega_2, \omega_3))$$

$$= \int I_E(\omega_1, \omega_2, \omega_3) f(\omega_1, \omega_2, \omega_3)\, d\,(\lambda_1 x \lambda_2 x \lambda_3 (\omega_1, \omega_2, \omega_3)) \quad (1.25)$$

$$= \int \{ \int I_E(\omega_1, \omega_2, \omega_3) \frac{f(\omega_1, \omega_2, \omega_3)}{g(\omega_1)} d\,(\lambda_2 x \lambda_3 (\omega_2, \omega_3))\} g(\omega_1)\, d\lambda(\omega_1)$$

$$= \int \mu_{.\,2\,3}(E \mid \omega_1) g(\omega_1)\, d\lambda_1(\omega_1) = \int \mu_{.\,2\,3}(E \mid \omega_1)\, d\mu_{1\,.\,.}(\omega_1)$$

where

$$\mu_{.\,2\,3}(E \mid \omega_1) = \int I_E(\omega_1, \omega_2, \omega_3) \frac{f(\omega_1, \omega_2, \omega_3)}{g(\omega_1)} d\,(\lambda_2 x \lambda_3 (\omega_2, \omega_3)) \quad (1.26)$$

$$= \int_{E_{\omega_1}} f(\omega_2, \omega_3 \mid \omega_1)\, d\,(\lambda_2 x \lambda_3 (\omega_2, \omega_3))$$

etc. Thus

$$I(\mu_{1\,2\,3} : \mu_{1\,.\,.} x \mu_{.\,2\,.} x \mu_{.\,.\,3}) =$$

$$\int f(\omega_1, \omega_2, \omega_3) \ln \frac{f(\omega_1, \omega_2, \omega_3)}{g(\omega_1) h(\omega_2) i(\omega_3)} d\,(\lambda_1 x \lambda_2 x \lambda_3 (\omega_1, \omega_2, \omega_3)) \quad (1.27)$$

$$= \int f(\omega_1, \omega_2, \omega_3) \ln \frac{f(\omega_1, \omega_2, \omega_3)}{f(\omega_1, \omega_2, \cdot\ ) i(\omega_3)} d\,(\lambda_1 x \lambda_2 x \lambda_3 (\omega_1, \omega_2, \omega_3)) +$$

$$+ \int f(\omega_1, \omega_2, \cdot\ ) \ln \frac{f(\omega_1, \omega_2, \cdot\ )}{g(\omega_1) h(\omega_2)} d\,(\lambda_1 x \lambda_2 (\omega_1, \omega_2)) =$$

$$= I(\mu_{1\,2\,3} : \mu_{1\,2\,.} x \mu_{.\,.\,3}) + I(\mu_{1\,2\,.} : \mu_{1\,.\,.} x \mu_{.\,2\,.}),$$

also

$$\int f(\omega_1, \omega_2, \omega_3)\, ln\, \frac{f(\omega_1, \omega_2, \omega_3)}{f(\omega_1, \omega_2, \cdot\ )\, i(\omega_3)}\, d(\lambda_1 x \lambda_2 x \lambda_3 (\omega_1, \omega_2, \omega_3))$$

$$= \int f(.\,, \omega_2, \omega_3)\, ln\, \frac{f(.\,, \omega_2, \omega_3)}{h(\omega_2)\, i(\omega_3)}\, d(\lambda_2 x \lambda_3 (\omega_2, \omega_3)) \tag{1.28}$$

$$+ \int f(\omega_1, \omega_2, \omega_3)\, ln\, \frac{f(\omega_1, \omega_2, \omega_3)}{\dfrac{f(\omega_1, \omega_2, .\,)\, f(.\,, \omega_2, \omega_3)}{h(\omega_2)}}\, d\,(\lambda_1 x \lambda_2 x \lambda_3 (\omega_1, \omega_2, \omega_3)).$$

Note that

$$\frac{f(\omega_1, \omega_2, .\,)\, f(.\,, \omega_2, \omega_3)}{h(\omega_2)} = f(\omega_1 | \omega_2)\, f(\omega_3 | \omega_2)\, h(\omega_2), \tag{1.29}$$

$$\int_{A_1 x A_2 x A_3} f(\omega_1 | \omega_2)\, f(\omega_3 | \omega_2)\, h(\omega_2)\, d(\lambda_1 x \lambda_2 x \lambda_3 (\omega_1, \omega_2, \omega_3))$$

$$= \int_{A_2} \mu_{1\,..}(A_1 | \omega_2)\, \mu_{..\,3}(A_3 | \omega_2)\, h(\omega_2)\, d\lambda_2(\omega_2) \tag{1.30}$$

$$= \mu_{1 x 3 | 2}(A_1 x A_2 x A_3)$$

and

$$\mu_{1 x 3 | 2}(E) = \int_E f(\omega_1 | \omega_2)\, f(\omega_3 | \omega_2)\, h(\omega_2)\, d(\lambda_1 x \lambda_2 x \lambda_3 (\omega_1, \omega_2, \omega_3)), \tag{1.31}$$

$$E \in \mathbf{A}_1 x \mathbf{A}_2 x \mathbf{A}_3,$$

$$\mu_{1x3|2}(G\,x\,\Omega_2) = \int_{G\,x\,\Omega_2} f(\omega_1|\omega_2)f(\omega_3|\omega_2)\,h(\omega_2)\,d(\lambda_1 x \lambda_2 x \lambda_3(\omega_1,\omega_2,\omega_3))$$

$$= \int_G \{\int f(\omega_1|\omega_2)f(\omega_3|\omega_2)\,h(\omega_2)\,d\lambda_2(\omega_2)\}\,d(\lambda_1 x \lambda_3(\omega_1,\omega_3)) \tag{1.32}$$

$$= \int_G g(\omega_1,\omega_3)\,d(\lambda_1 x \lambda_3(\omega_1,\omega_3)) = \mu_{1x3|.}(G),\, G \in \mathbf{A}_1 x \mathbf{A}_3$$

where

$$g(\omega_1,\omega_3) = \int f(\omega_1|\omega_2)f(\omega_3|\omega_2)\,h(\omega_2)\,d\lambda_2(\omega_2), \tag{1.33}$$

and $\mu_{1x3|2}(\Omega_1 x F) = \int_{\Omega_1 x F} f(\omega_1|\omega_2)\,f(\omega_3|\omega_2)\,h(\omega_2)\,d(\lambda_1 x \lambda_2 x \lambda_3(\omega_1,\omega_2,\omega_3))$

$$= \int_F f(.\,\omega_2,\omega_3)\,d(\lambda_2 x \lambda_3(\omega_2,\omega_3)) = \mu_{.\,23}(F),\, F \in A_2 x A_3,$$

similarly $\mu_{1x3|2}(B\,x\,\Omega_3) = \mu_{12.}(B),\, B \in \mathbf{A}_1 x \mathbf{A}_2.$

It may be verified that $\mu_{1x3|2}(\Omega_1 x \Omega_2 x \Omega_3) = 1$.

We may also write

$$\int f(\omega_1,\omega_2,\omega_3)\,ln \frac{f(\omega_1,\omega_2,\omega_3)}{\dfrac{f(\omega_1,\omega_2,.)f(.,\omega_2,\omega_3)}{h(\omega_2)}}\,d(\lambda_1 x \lambda_2 x \lambda_3(\omega_1,\omega_2,\omega_3)) \tag{1.34}$$

$$= \int \{\int f(\omega_1,\omega_3|\omega_2)\,ln\,\frac{f(\omega_1,\omega_3|\omega_2)}{f(\omega_1|\omega_2)f(\omega_3|\omega_2)}\,d(\lambda_1 x \lambda_3(\omega_1,\omega_3))\}\,h(\omega_2)\,d\lambda_2(\omega_2)$$

and (1.28) as

$$I(\mu_{123} : \mu_{12.} \, x \, \mu_{..3}) = I(\mu_{.23} : \mu_{.2.} \, x \, \mu_{..3}) + I(\mu_{123} : \mu_{1x3|2}) \,. \tag{1.35}$$

Extensions to four and more variables are relatively straight forward.

Note that the convexity property insures the following inequalities

$$\int f(\omega_1, \omega_2, \omega_3) \, ln \, \frac{f(\omega_1, \omega_2, \omega_3)}{g(\omega_1) \, h(\omega_2) \, i(\omega_3)} \, d(\lambda_1 \, x \, \lambda_2 \, x \, \lambda_3 (\omega_1, \omega_2, \omega_3)) \tag{1.36}$$

$$\geq \int f(\omega_1, \omega_2, .) \, ln \, \frac{f(\omega_1, \omega_2, .)}{g(\omega_1) \, h(\omega_2)} \, d(\lambda_1 \, x \, \lambda_2 (\omega_1, \omega_2)),$$

recall that

$$f(\omega_1, \omega_2, .) = \int f(\omega_1, \omega_2, \omega_3) \, d\lambda_3 (\omega_3) \, , \, 1 = \int i(\omega_3) \, d\lambda_3 (\omega_3) \, ; \tag{1.37}$$

$$\int f(\omega_1, \omega_2, \omega_3) \, ln \, \frac{f(\omega_1, \omega_2, \omega_3)}{f(\omega_1, \omega_2, .) \, i(\omega_3)} \, d(\lambda_1 \, x \, \lambda_2 \, x \, \lambda_3 (\omega_1, \omega_2, \omega_3)) \tag{1.38}$$

$$\geq \int f(. \, , \omega_2, \omega_3) \, ln \, \frac{f(. \, , \omega_2, \omega_3)}{h(\omega_2) \, i(\omega_3)} \, d(\lambda_2 \, x \, \lambda_3 (\omega_2, \omega_3)),$$

$$\int_E f(\omega_1, \omega_2, \omega_3) \, ln \, \frac{f(\omega_1, \omega_2, \omega_3)}{\dfrac{f(\omega_1, \omega_2, .) \, f(. \, , \omega_2, \omega_3)}{h(\omega_2)}} \, d(\lambda_1 \, x \, \lambda_2 \, x \, \lambda_3 (\omega_1, \omega_2, \omega_3)) \tag{1.39}$$

$$\geq \mu_{123}(E) \, ln \, \frac{\mu_{123}(E)}{\mu_{1x3|2}(E)} \, , \, E \in \mathbf{A}_1 \, x \, \mathbf{A}_2 \, x \, \mathbf{A}_3 \, .$$

(See Table 3.3, p. 165, Table 3.5, p. 167, Kullback (1959).)

We may now rewrite some of the preceding expressions as relations among discrimination information values in a notation which is reasonably obvious. Thus (1.27) becomes

$$I(\omega_1, \omega_2, \omega_3) = I((\omega_1, \omega_2), \omega_3) + I(\omega_1, \omega_2); \tag{1.40}$$

(1.28) and (1.34) become

$$I((\omega_1, \omega_2), \omega_3) = I(\omega_2, \omega_3) + I(\omega_1, \omega_3 \mid \omega_2), \tag{1.41}$$

where $I(\omega_1, \omega_3 \mid \omega_2)$ is the expected value of the conditional value $I(\omega_1, \omega_3 \mid \omega_2 = \omega_2')$ (see (2.17) - (2.22) in Chapter II). The discrimination information values are non-negative and the inequalities (1.36), (1.38) are also derivable from (1.40), (1.41) as

$$I(\omega_1, \omega_2, \omega_3) \geq I(\omega_1, \omega_2) \tag{1.42}$$

$$I((\omega_1, \omega_2), \omega_3) \geq I(\omega_2, \omega_3). \tag{1.43}$$

Note that $I(\omega_1, \omega_2)$, $I(\omega_1, \omega_2, \omega_3)$, etc. represent discrimination information measures between a hypothesis of dependence and one of independence of the random variables or set of random variables.

We may of course apply the results in Chapter II and Chapter III to the case of discrimination information measures for hypotheses of dependence and independence. As an application of Theorem 2.1 of Chapter II we state the following. Suppose that the random variable $x_1 = x_1(\omega_1)$ is a measurable function of ω_1, then

$$I(\omega_1, \omega_2) \geq I(x_1, \omega_2). \tag{1.44}$$

The inequalities (1.42) and (1.43) are also special cases of Theorem 2.1 of Chapter II, and more generally

$$I((\omega_i, i \in T_1), (\omega_j, j \in T_2)) \geq I(\omega_i, i \in T_1'), (\omega_j, j \in T_2')) \tag{1.45}$$

where T_1 and T_2 are disjoint index sets and $T_1' \subseteq T_1$, $T_2' \subseteq T_2$.

As an application of (2.19) or (3.26) in Chapter III consider the following. For notational convenience let use denote one of the random variables as ω' and let $\omega = (\omega_1, \omega_2, \ldots)$, then

$$\lim_{n \to \infty} I((\omega_1, \omega_2, \ldots, \omega_n), \omega') = I(\omega, \omega') . \tag{1.46}$$

The random variables $\omega', \omega_1, \omega_2, \ldots$ take their values in the measurable spaces $(\Omega', \mathbf{A}'), (\Omega_1, \mathbf{A}_1), (\Omega_2, \mathbf{A}_2), \ldots$ and the random variables $\omega', (\omega_1, \omega_2, \ldots, \omega_n)$ take values in the measurable product space $(\Omega' x\, \Omega_1 x\, \Omega_2 x \ldots x\, \Omega_n, \mathbf{A}' x\, \mathbf{A}_1 x \ldots x\, \mathbf{A}_n)$. If we set $\beta_n = \mathbf{A}' x\, \mathbf{A}_1 x \ldots x\, \mathbf{A}_n$, $\beta = \beta_\infty = \mathbf{A}' x\, \mathbf{A}_1 x \ldots x\, \mathbf{A}_n \ldots$ then $\beta_n \uparrow \beta$ and the desired result (1.46) is then a consequence of (2.19) or (3.26) in Chapter III.

As an application of Theorem 1.1 of Chapter III consider the following. Let $\omega_1, \omega_2, \ldots, \omega$ and ω' be random variables taking values in the measurable spaces $(\Omega, \mathbf{A}), (\Omega', \mathbf{A}')$ respectively. Then if the joint distribution of (ω_i, ω') converges to the joint distribution of (ω, ω')

$$\liminf_{n \to \infty} I(\omega_n, \omega') \geq I(\omega, \omega'). \tag{1.47}$$

2. Markov Processes

Let us now consider the discrimination information between two Markov processes. Suppose the measure spaces $(\Omega_t, \mathbf{A}_t, \lambda_t)$, $t = 1, 2, \ldots$ are all copies of the same measure space and consider t as denoting time. Suppose there are two generalized densities

$$f_1(\omega_1, \omega_2, \ldots, \omega_n), \quad f_2(\omega_1, \omega_2, \ldots, \omega_n) \tag{2.1}$$

defined on the product measure space

$$(\Omega_1 x \Omega_2 x \ldots x \Omega_n, \mathbf{A}_1 x \mathbf{A}_2 x \ldots x \mathbf{A}_n, \lambda_1 x \lambda_2 x \ldots x \lambda_n)$$

and which satisfy a Markovity condition, that is, are such that

$$f_i(\omega_1, \omega_2, \ldots, \omega_n) = f_i(\omega_1) f_i(\omega_2|\omega_1) f_i(\omega_3|\omega_2) \ldots f_i(\omega_n|\omega_{n-1}), \; i = 1, 2 \tag{2.2}$$

and the conditional densities do not depend on the time origin,

$$f_i(\omega_2|\omega_1) = f_i(\omega_3|\omega_2) = \ldots = f_i(\omega_n|\omega_{n-1}), \; i = 1, 2. \tag{2.3}$$

For convenience, let us set

$$\beta_n = \mathbf{A}_1 x \mathbf{A}_2 x \ldots x \mathbf{A}_n, \; \beta = \beta_\infty \tag{2.4}$$

and

$$I(1:2; \beta_n) = \int f_1(\omega_1, \ldots, \omega_n) \, ln \, \frac{f_1(\omega_1, \ldots, \omega_n)}{f_2(\omega_1, \ldots, \omega_n)} \, d(\lambda_1 x \ldots x \lambda_n (\omega_1, \ldots, \omega_n)) \tag{2.5}$$

then using (2.2) and (2.3)

$$\begin{aligned} I(1:2; \beta_n) &= \int f_1(\omega_1) \, ln \, \frac{f_1(\omega_1)}{f_2(\omega_1)} \, d\lambda_1(\omega_1) \\ &+ (n-1) \int f_1(\omega_1) \, d\lambda_1(\omega_1) \int f_1(\omega_2|\omega_1) \, ln \, \frac{f_1(\omega_2|\omega_1)}{f_2(\omega_2|\omega_1)} \, d\lambda_2(\omega_2) \\ &= I(1:2; \omega_1) + (n-1) \, I(1:2; \omega_2|\omega_1), \end{aligned} \tag{2.6}$$

where $I(1:2; \omega_2|\omega_1)$ is the expected value of the conditional value $I(1:2; \omega_2| \omega_1 = \omega_1')$. Using (2.19) or (3.26) in Chapter III (see the discussion following (1.46))

$$\lim_{n\to\infty} I(1:2;\beta_n) = I(1:2;\beta) = I(1:2;\omega_1) + \lim_{n\to\infty} (n-1) I(1:2;\omega_2|\omega_1) \tag{2.7}$$

$$I(1:2;\omega_2|\omega_1) = \lim_{n\to\infty} \frac{1}{n-1} I(1:2;\beta_n). \tag{2.8}$$

We shall now derive a result similar to (2.7) in terms of the notation and concepts in Chapters II and III. Suppose there is a sequence of sub-sigma-algebras of $(\Omega, \mathbf{A})$ $\beta_0, \beta_1, \beta_2, \ldots, \beta_n, \ldots$ and let us denote

$$\beta^j = \vee_{i=1}^j \beta_i. \tag{2.9}$$

We note that

$$\beta^0 \subset \beta^1 \subset \beta^2 \ldots \subset \beta^n \subset \ldots$$

and as in (2.1) of Chapter III we have $\beta^n \uparrow \beta$. As in Theorem 2.3 of Chapter II

$$\begin{aligned} I(\beta^n) &= \int E^{\beta^n} Z_t \, ln \frac{E^{\beta^n} Z_t}{E^{\beta^n} Z} dP \\ &= \int E^{\beta^{n-1}} Z_t \, ln \frac{E^{\beta^{n-1}} Z_t}{E^{\beta^{n-1}} Z} dP + \int E^{\beta^n} Z_t \, ln \frac{E^{\beta^n} Z_t / E^{\beta^{n-1}} Z_t}{E^{\beta^n} Z / E^{\beta^{n-1}} Z} dP \\ &= I(\beta^{n-1}) + I(\beta^n | \beta^{n-1}), \end{aligned} \tag{2.10}$$

hence, by iterating

$$I(\beta^n) = I(\beta^1) + I(\beta^2|\beta^1) + \ldots + I(\beta^n|\beta^{n-1}). \tag{2.11}$$

Suppose that the sequence of sub-sigma-algebras $\{\beta n\}$ satisfies a stationary Markov property, that is,

$$I(\beta^n|\beta^{n-1}) = \ldots = I(\beta^j|\beta^{j-1}) = \ldots = I(\beta^2|\beta^1), \tag{2.12}$$

then

$$I(\beta^n) = I(\beta^1) + (n-1)I(\beta^2|\beta^1). \tag{2.13}$$

Hence applying (2.19) or (3.26) in Chapter III

$$\lim_{n\to\infty} I(\beta^n) = I(\beta) = I(\beta^1) + \lim_{n\to\infty} (n-1)\, I(\beta^2|\beta^1). \tag{2.14}$$

As an application of the preceding results and Theorem 3.1 in Chapter III let us determine the discrimination information for covariance stationary normal Markov processes. We shall write the process as $\{X(t), t \in T\}$ and the random variable $X(t_i)$ when $t = t_i$ will represent the ω_i of the previous discussion. We shall consider the stationary normal Markov processes defined by the conditional densities (for a stationary Markov chain see Kullback (1966, 1967a))

$$f_a(x(t_{i+1})|x(t_i)) = \frac{1}{(2\pi(1-\rho^2))^{\frac{1}{2}}}\, exp\{-\tfrac{1}{2}(x(t_{i+1}) - m_a - \rho(x(t_i) - m_a))^2/ (1-\rho)^2\}, a = 1, 2, \tag{2.15}$$

where $\rho = \mathrm{cov}(x(t_i), x(t_{i+1}))$, and the marginal densities are

$$f_a(x(t)) = \frac{1}{(2\pi)^{\frac{1}{2}}}\, exp\{-\tfrac{1}{2}(x(t) - m_a)^2\},\ a = 1, 2. \tag{2.16}$$

For convenience and without limiting the generality let us take

$$H_1 : E_1(x(t)) = m_1 = m,\ H_2 : E_2(x(t)) = m_2 = 0. \tag{2.17}$$

It is found that (see example 2.1, page 14, Kullback (1959))

$$\int f_1(x(t_{i+1})|x(t_i))\, ln\, \frac{f_1(x(t_{i+1})|x(t_i))}{f_2(x(t_{i+1})|x(t_i))}\, dx(t_{i+1}) \tag{2.18}$$

$$= (m + \rho(x(t_i) - m) - \rho x(t_i))^2 / 2(1 - \rho^2)$$

and hence what had been designated $I(1:2;\omega_2|\omega_1)$ in (2.6) is

$$I(1:2;\omega_2|\omega_1) = E\{(m + \rho(x(t_i) - m) - \rho(x(t_i))^2/2(1-\rho^2)\}$$

$$= \frac{m^2(1-\rho)^2}{2(1-\rho^2)} = \frac{m^2(1-\rho)}{2(1+\rho)} . \tag{2.19}$$

Using (2.16), (2.17) it is found that (see example 2.1, page 14, Kullback (1959)),

$$\int f_1(x(t))\, ln\, \frac{f_1(x(t))}{f_2(x(t))}\, dx(t) = m^2/2 \tag{2.20}$$

that is, in the notation of (2.6)

$$I(1:2;\omega_1) = m^2/2 . \tag{2.21}$$

Ornstein-Uhlenbeck Process

In particular let us take the Ornstein-Uhlenbeck process for which the correlation function is

$$R(s,t) = cov(x(s), x(t)) = exp\{-\beta|t-s|\},\ s,\ t \in T \tag{2.22}$$

where we note that the variables here have unit variances. Hence the value of ρ in (2.15) is $\rho = exp\{-\beta(t_{i+1} - t_i)\}$, so that if we consider the processes during the time $0 \le t \le T$ and set $t_{i+1} - t_i = T/(n-1) = \Delta t$, then

$$1 - \rho = \beta\Delta t + 0(\Delta t)^2, 1 + \rho = 2 - \beta\Delta t + 0(\Delta t)^2, \tag{2.23}$$

and

$$I(1:2;\omega_2|\omega_1) = \frac{m^2}{2}\,\frac{\beta\Delta t + 0(\Delta t)^2}{2 - \beta\Delta t + 0(\Delta t)^2} \tag{2.24}$$

$$(n-1)\,I(1:2;\omega_2|\omega_1) = \frac{m^2}{2}\,\frac{\beta T + 0(\Delta t)}{2 - \dfrac{\beta T}{n-1} + 0(\Delta t)^2}\,. \tag{2.25}$$

Thus for T finite, as $\Delta t \to 0$, that is, $n \to \infty$

$$\lim_{n\to\infty} (n-1)I(1:2;\omega_2|\omega_1) = \frac{m^2}{2}\,\frac{\beta T}{2} \tag{2.26}$$

and using (2.7), (2.21), and (2.26)

$$I(1:2;\beta) = I(1:2;X(t), 0 \le t \le T) = \frac{m^2}{2} + \frac{m^2}{2}\,\frac{\beta T}{2} = \frac{m^2}{2}\left(1 + \frac{\beta T}{2}\right). \tag{2.27}$$

Let us now assume that the mean values depend on the time, that is, instead of (2.17)

$$H_1 : E_1(X(t)) = m(t),\ H_2 : E_2(X(t)) = 0\,, \tag{2.28}$$

where we shall assume that the function $m(t)$ has the properties needed in the following. Instead of (2.18), now

$$\int f_1(x(t_{i+1})|x(t_i))\,ln\,\frac{f_1(x(t_{i+1})|x(t_i))}{f_2(x(t_{i+1})|x(t_i))}\,dx(t_{i+1}) \tag{2.29}$$

$$= (m(t_{i+1}) + \rho(x(t_i) - m(t_i)) - \rho x(t_i))^2/2(1-\rho^2)$$

and instead of (2.19),

$$I(1:2;\omega_2|\omega_1) = E_1\{(m(t_{i+1}) + \rho(x(t_i) - m(t_i)) - \tag{2.30}$$

$$\rho x(t_i))^2/2(1-\rho^2)\} = (m(t_{i+1}) - \rho m(t_i))^2/2(1-\rho^2)\,.$$

For $t_{i+1} = t_i + \Delta t$, and the same approximations as in (2.23), instead of (2.24),

$$I(1:2; x(t_{i+1})|x(t_i)) =$$

$$= \frac{(m(t_i) + \Delta t m'(t_i) + 0(\Delta t)^2 - m(t_i) + \beta m(t_i)\Delta t + 0(\Delta t)^2)^2}{2(\beta\Delta t + 0(\Delta t)^2)(2 - \beta\Delta t + 0(\Delta t)^2)}$$

$$= \frac{(m'(t_i)\Delta t + \beta m(t_i)\Delta t + 0(\Delta t)^2)^2}{2(2\beta\Delta t + 0(\Delta t)^2)} \tag{2.31}$$

$$= \frac{(m'(t_i) + \beta m(t_i) + 0(\Delta t))^2\,\Delta t}{4\beta + 0(\Delta t)}$$

$$= \frac{(m'(t_i))^2 + 2\beta m(t_i)m'(t_i) + \beta^2(m(t_i))^2 + 0(\Delta t)}{4\beta + 0(\Delta t)}\,\Delta t\,,$$

instead of (2.21)

$$I(1:2;\omega_1) = (m(0))^2/2 \tag{2.32}$$

where $0 \le t \le T$ in (2.31) and (2.32).

Hence now, instead of (2.27), the conditional densities in (2.2) are not identical,

$$\lim_{n\to\infty} I(1:2;\beta_n) = I(1:2;\beta) = I(1:2;X(t), 0 \le t \le T)$$

$$= \frac{m^2(0)}{2} + \lim_{\Delta t\to 0} \sum \frac{(m'(t_i))^2 + 2\beta m(t_i)m'(t_i) + \beta^2(m(t_i))^2}{4\beta}\,\Delta t$$

$$= \frac{m^2(0)}{2} + \tfrac{1}{4}\int_0^T [\frac{1}{\beta}(m'(t))^2 + \beta(m(t))^2]\,dt + \tfrac{1}{2}\int_0^T m(t)\,m'(t)\,dt \tag{2.33}$$

$$= \frac{m^2(0)}{2} + \tfrac{1}{4}\int_0^T [\frac{1}{\beta}(m'(t))^2 + \beta(m(t))^2]\,dt + \tfrac{1}{4}(m(t))^2\Big|_0^T$$

$$= \tfrac{1}{4}\int_0^T [\frac{1}{\beta}(m'(t))^2 + \beta(m(t))^2]\,dt + \frac{(m(0))^2 + (m(T))^2}{4}.$$

Note that for m (t) = m, m ' (t) = 0, the value just derived reduces to that in (2.27).

On pages 84, 94-5 of Kullback (1959) there is exhibited for members of an exponential family a relation between the difference of two minimum discrimination information statistics and the logarithm of likelihood ratios. We shall use this result to obtain the log-likelihood ratio of the Ornstein-Uhlenbeck process as well as other processes considered later. We are indebted to Dr. Edward Melnick who used this procedure for other stochastic processes. (Melnick (1973), Melnick and Kullback (1973)).

The minimum discrimination information statistic in this case may be obtained from the discrimination information by replacing parameters by best unbiased estimates. Thus using the observed process X (t) as the statistic we have from (2.33)

$$\hat{\mathrm{I}}(*:2;\beta) = \tfrac{1}{4}\int_0^T [\frac{1}{\beta}(X'(t))^2 + \beta(X(t))^2]\,dt + \frac{(X(0))^2 + (X(T))^2}{4} \tag{2.34}$$

and

$$\hat{\mathrm{I}}(*:1;\beta) = \tfrac{1}{4}\int_0^T [\frac{1}{\beta}(X'(t) - m'(t))^2 + \beta(X(t) - m(t))^2]\,dt + \frac{(X(0) - m(0))^2 + (X(T) - m(T))^2}{4} \tag{2.35}$$

where in (2.34) and (2.35) the parameters corresponding to H_1 and H_2 in (2.28) have been respectively replaced in (2.33). Hence

$$\hat{\mathrm{I}}(*:2;\beta) - \hat{\mathrm{I}}(*:1;\beta) = \tfrac{1}{4}\int_0^T [\frac{2}{\beta}(X'(t)\,m'(t) + 2\beta X(t)\,m(t)]\,dt$$

$$- \tfrac{1}{4} \int_0^T [\frac{1}{\beta} (m'(t))^2 + \beta (m(t))^2] \, dt + \tag{2.36}$$

$$+ \frac{2X(0)m(0) + 2X(T)m(T) - (m(0))^2 - (m(T))^2}{4} .$$

If we write

$$\frac{1}{2\beta} \int_0^T m'(t) X'(t) \, dt = \frac{X(T) m'(T) - X(0) m'(0)}{2\beta} - \frac{1}{2\beta} \int_0^T X(t) m''(t) \, dt \tag{2.37}$$

we finally have for $\ell(X(t))$ the log-likelihood ratio

$$\hat{\mathrm{I}}(*:2;\beta) - \hat{\mathrm{I}}(*:1;\beta) = \ell(X(t)) = ln \frac{f_1(X(t), 0 \le t \le T)}{f_2(X(t), 0 \le t \le T)} = \tag{2.38}$$

$$\frac{m(0)X(0) + m(T)X(T)}{2} - \frac{(m(T))^2 + (m(0))^2}{4} + \frac{X(T)m'(T) - X(0)m'(0)}{2\beta}$$

$$- \frac{\beta}{4} \int_0^T m^2(t) \, dt - \frac{1}{4\beta} \int_0^T (m'(t))^2 \, dt + \tfrac{1}{2} \int_0^T X(t) [\beta m(t) - \frac{1}{\beta} m''(t)] \, dt .$$

Note that for $m(t) = m, m'(t) = m''(t) = 0$ the result in (2.38) yields for the log-likelihood ratio

$$\ell(X(t)) = \frac{m}{2} (X(0) + X(T)) - \frac{m^2}{4} (2 + \beta T) + \frac{m\beta}{2} \int_0^T X(t) \, dt$$

a result given in Grenander (1950, p. 228). For a computation example see Shumway (1965).

In Striebel (1959) using a different approach there was also considered the regression case where $m(t) = \sum_{i=1}^{s} k_i \Phi_i(t)$, where the $\Phi_i(t)$ are known continuous functions and the k_i are parameters. For another method of deriving the final value in (2.33) see Lilliefors (1964) and for (2.38) see Yaglom in Rosenblatt (1963, p. 330).

The Wiener Process

As another illustration we consider the Wiener process for the time interval $a \le t \le b$. For this process the joint density of $x(s)$ and $x(t)$ for $s < t$ is given by

$$f_\alpha(x(s))\, f_\alpha(x(t)\,|\,x(s)) = \frac{1}{(2\pi s)^{\frac{1}{2}}} \exp\left\{-\frac{(x(s) - m_\alpha(s))^2}{2s}\right\} \times \tag{2.39}$$

$$\frac{1}{(2\pi(t-s))^{\frac{1}{2}}} \exp\left\{-\frac{((x(t) - m_\alpha(t)) - (x(s) - m_\alpha(s)))^2}{2(t-s)}\right\}, \alpha = 1, 2$$

with the covariance function $\operatorname{cov}(x(s), x(t)) = \sigma(s,t) = \min(s,t)$. The Wiener process is also a Markov process. For convenience let us take

$$H_1 : E_1(x(t)) = m_1(t) = m(t),\ H_2 : E_2(x(t)) = m_2 \equiv 0\,. \tag{2.40}$$

With the densities in (2.39), it is found that (see example 2.1, p. 14, Kullback (1959))

$$\int f_1(x(t)\,|\,x(s))\, \ln \frac{f_1(x(t)\,|\,x(s))}{f_2(x(t)\,|\,x(s))}\, dx(t) = \frac{(m(t) + (x(s) - m(s)) - x(s))^2}{2(t-s)} \tag{2.41}$$

and in the notation of (2.6)

$$I(1:2\,;x(t)\,|\,x(s)) = E_1\,\frac{(m(t) + (x(s) - m(s)) - x(s))^2}{2(t-s)} \tag{2.42}$$

$$= \frac{(m(t) - m(s))^2}{2(t-s)},$$

$$I(1:2\,;\omega_1) = (m(a))^2/2a. \tag{2.43}$$

Now consider the limit relations

$$t - s = \Delta t,\ \Delta t = (b - a)/(n - 1) \tag{2.44}$$

then (see (2.33))

$$\lim_{n\to\infty} I(1:2\,;\beta_n) = I(1:2\,;\beta) = I(1:2\,;X(t),\ a \le t \le b)$$

$$= \tfrac{1}{2}\,\frac{(m(a))^2}{a} + \lim_{\Delta t\to 0}\ \tfrac{1}{2}\sum \frac{(m(t+\Delta t) - m(t))^2}{\Delta t} \tag{2.45}$$

$$= \tfrac{1}{2}\,\frac{(m(a))^2}{a} + \tfrac{1}{2}\int_a^b \frac{(d\,m(t))^2}{dt}.$$

Note that if $m(t) = m,$ then (2.45) becomes

$$I(1:2\,;X(t),\ a \le t \le b) = (m)^2/2a. \tag{2.46}$$

For another method of deriving the discrimination information value in (2.45) see Lilliefors (1964). We remark that if $m(t)$ in (2.45) has a derivative then

$$\int_a^b \frac{(d\,m(t))^2}{dt} = \int_a^b (m'(t))^2\,dt. \tag{2.47}$$

To derive the log-likelihood ratio in this case we follow a procedure similar to that used in (2.34) - (2.38). Again using the observed process $X(t)$ as the statistic we have from (2.45)

$$\hat{I}(*:2;\beta) = \tfrac{1}{2}\,\frac{(X(a))^2}{a} + \tfrac{1}{2}\int_a^b \frac{(dX(t))^2}{dt} \tag{2.48}$$

$$\hat{I}(*:1;\beta) = \tfrac{1}{2}\,\frac{(X(a) - m(a))^2}{a} + \tfrac{1}{2}\int_a^b \frac{(d(X(t) - m(t)))^2}{dt} \tag{2.49}$$

where in (2.48) and (2.49) the parameters corresponding to H_1 and H_2 in (2.40) have been respectively replaced in (2.45). Hence

$$\hat{I}(*:2;\beta) - \hat{I}(*:1;\beta) = \frac{2X(a)m(a) - (m(a))^2}{2a} + \int_a^b \frac{(dX(t)\,dm(t)}{dt} - \tfrac{1}{2}\int_a^b \frac{(dm(t))^2}{dt}\,, \tag{2.50}$$

that is

$$\ell\,(X(t)) = ln\,\frac{f_1(X(t), a \le t \le b)}{f_2(X(t), a \le t \le b)} \tag{2.51}$$

is given by the right-hand side of (2.50). In Lilliefors (1964), Par. 9.4 for the case $a = 0,\ b = 1$ and assuming that $m(t)$ is twice differentiable, the value corresponding to (2.50) was derived by another approach. See also Yaglom in Rosenblatt (1963, p. 329).

Different Covariance Functions

We shall now consider a case in which the means are the same under both hypotheses but the covariance functions are different. Consider the stationary Gaussian Markov processes with covariance functions

$$H_1 : B_1(t) = \frac{1}{\beta}\, exp\,(-\beta|t|),\ H_2 : B_2(t) = \frac{1}{\alpha}\, exp\,(-\alpha|t|). \tag{2.52}$$

The corresponding spectral densities are

$$H_1 : f_1(\lambda) = 1/\pi\,(\lambda^2 + \beta^2),\ H_2 : f_2(\lambda) = 1/\pi\,(\lambda^2 + \alpha^2), \tag{2.53}$$

where (see Doob (1953, p. 522))

$$f_n(\lambda) = \frac{1}{2\pi}\int_{-\infty}^{\infty} e^{-i\lambda t} B_n(t)\,dt,\ i = (-1)^{\frac{1}{2}},\ n = 1, 2.$$

In Rosenblatt (1963, p. 337) Yaglom gives for this case

$$f(x(t)) = \frac{P_1(dx)}{P_2(dx)} = \left(\frac{\beta}{\alpha}\right)^{\frac{1}{2}} exp\,\{-[\frac{\beta-\alpha}{4}(x^2(0) + x^2(T) - 2T) + \\ + \frac{\beta^2-\alpha^2}{4}\int_0^T x^2(t)\,dt]\} \tag{2.54}$$

hence, taking the logarithm of (2.54) and finding the expected value under H_1, we have

$$I(1:2; X(t), 0 \leq t \leq T) = \tfrac{1}{2}\, ln\,\frac{\beta}{\alpha} + \frac{\alpha-\beta}{2\beta} + T\,\frac{(\alpha-\beta)^2}{4\beta}, \tag{2.55}$$

and the rate of discrimination information is (see Shumway (1965, p. 48), also (2.6) of Chapter VI)

$$\bar{I} = \lim_{T\to\infty} \frac{1}{T} I(1:2; X(t), -T \leq t \leq T) \tag{2.56}$$

$$= \frac{1}{2\pi}\int_{-\infty}^{\infty} (\frac{f_1(\lambda)}{f_2(\lambda)} - 1 - ln\,\frac{f_1(\lambda)}{f_2(\lambda)})\,d\lambda$$

$$= \frac{1}{2\pi} \int_{-\infty}^{\infty} \left(\frac{\alpha^2 - \beta^2}{\lambda^2 + \beta^2} - ln \frac{\lambda^2 + \alpha^2}{\lambda^2 + \beta^2} \right) d\lambda$$

$$= \frac{\alpha^2 - \beta^2}{2\beta} - \frac{1}{2\pi} \int_{-\infty}^{\infty} ln \frac{\lambda^2 + \alpha^2}{\lambda^2 + \beta^2} d\lambda$$

From tables of integrals it is found that

$$\int ln\,(\alpha^2 + \lambda^2)\, d\lambda = \lambda\, ln\,(\alpha^2 + \lambda^2) - 2\lambda + 2\alpha\, tan^{-1} \frac{\lambda}{\alpha} \tag{2.57}$$

from which it follows that

$$\int_{-\infty}^{\infty} ln \frac{\alpha^2 + \lambda^2}{\beta^2 + \lambda^2} d\lambda = 2\pi\,(\alpha - \beta), \tag{2.58}$$

so that from (2.56)

$$\bar{I} = \frac{(\alpha - \beta)^2}{2\beta} . \tag{2.59}$$

We note that from (2.55)

$$I(1:2; X(t), -T \le t \le T) = ln \frac{\beta}{\alpha} + \frac{\alpha - \beta}{\beta} + T \frac{(\alpha - \beta)^2}{2\beta} \tag{2.60}$$

and verify that

$$\lim_{T \to \infty} \frac{1}{T} I(1:2; X(t), -T \le t \le T) = \frac{(\alpha - \beta)^2}{2\beta} . \tag{2.61}$$

For this same case Striebel (1959) found the inverse of the covariance matrices of the joint distributions and derived

$$\frac{dP_{\sigma_1, \beta_1}}{dP_{\sigma_2, \beta_2}} = \left(\frac{\beta_1}{\beta_2}\right)^{\frac{1}{2}} exp\{ - \frac{1}{2K} ((\beta_1 - \beta_2)(x^2(0) + x^2(T) - KT) + + (\beta_1^2 - \beta_2^2) \int_0^T x^2(t)\,dt\} \tag{2.62}$$

for $2\sigma^2\beta = 2\sigma_1^2\beta_1 = 2\sigma_2^2\beta_2 = K$. The case just discussed corresponds to $K = 2$.

To use the relation for the discrimination information values and densities corresponding to the Markov property we use (see Kullback (1959, p. 14))

$$h_i(y|x) = \frac{1}{\sigma_{iy}(2\pi(1-\rho_i^2))^{\frac{1}{2}}} exp\left\{\frac{(y - \mu_{iy} - \beta_i(x - \mu_{ix}))^2}{2\sigma_{iy}^2(1 - \rho_i^2)}\right\}. \tag{2.63}$$

In this case we take $\mu_{iy} = 0 = \mu_{ix}$, and

$$I(1:2;Y|X = x) = \tfrac{1}{2} ln \frac{\sigma_{2y}^2(1-\rho_2^2)}{\sigma_{1y}^2(1-\rho_1^2)} - \tfrac{1}{2} + \frac{\sigma_{1y}^2(1-\rho_1^2)}{2\sigma_{2y}^2(1-\rho_2^2)} + \frac{(\beta_2 x - \beta_1 x)^2}{2\sigma_{2y}^2(1-\rho_2^2)} \tag{2.64}$$

and taking the expected value of (2.64) over all x

$$I(1:2;Y|X) = \tfrac{1}{2} ln \frac{\sigma_{2y}^2(1-\rho_2^2)}{\sigma_{1y}^2(1-\rho_1^2)} - \tfrac{1}{2} + \frac{\sigma_{1y}^2(1-\rho_1^2)}{2\sigma_{2y}^2(1-\rho_2^2)} + \frac{(\beta_2 - \beta_1)^2\sigma_{1x}^2}{2\sigma_{2y}^2(1-\rho_2^2)}. \tag{2.65}$$

For the case under consideration, the regression coefficient $\beta_i = \rho_i \sigma_{iy}/\sigma_{ix} = \rho_i$, since we take $\sigma_{iy} = \sigma_{ix}$ (we are using Y for $X(t + \Delta t)$ and X for $X(t)$). Using Yaglom's notation as in (2.54) we have

$$\sigma_1^2 = 1/\beta, \quad \sigma_2^2 = 1/\alpha \tag{2.66}$$

and using the notation in (2.23)

$$1 - \rho_1 = \beta\Delta t + 0(\Delta t)^2, 1 + \rho_1 = 2 - \beta\Delta t + 0(\Delta t)^2, \tag{2.67}$$

$$1 - \rho_2 = \alpha\Delta t + 0(\Delta t)^2, 1 + \rho_2 = 2 - \alpha\Delta t + 0(\Delta t)^2$$

we get, using ω_2 for $X(t + \Delta t)$ and ω_1 for $X(t)$, to within terms of order Δt,

$$I(1:2;\omega_2|\omega_1) = \tfrac{1}{2}\, ln\, \frac{\beta}{\alpha} + \tfrac{1}{2}\, ln\, \frac{\alpha\Delta t(2 - \alpha\Delta t)}{\beta\Delta t(2 - \beta\Delta t)}$$

$$- \tfrac{1}{2} + \frac{\frac{1}{\beta}\beta\Delta t(2 - \beta\Delta t)}{2\frac{1}{\alpha}\alpha\Delta t(2 - \alpha\Delta t)} + \frac{(1 - \alpha\Delta t - 1 + \beta\Delta t)^2}{\beta\frac{2}{\alpha}\alpha\Delta t(2 - \alpha\Delta t)} \tag{2.68}$$

$$= \frac{(\beta - \alpha)^2}{4\beta}\Delta t.$$

Hence

$$I(1:2;X(t), 0 \le t \le T) = I(1:2;\omega_1) + \lim_{n\to\infty}(n-1)\, I(1:2;\omega_2|\omega_1) \tag{2.69}$$

and since $\Delta t = T/(n-1)$ we have

$$I(1:2;X(t), 0 \le t \le T) = \tfrac{1}{2}\left(ln\, \frac{\beta}{\alpha} - 1 + \frac{\alpha}{\beta}\right) + \frac{(\beta - \alpha)^2}{4\beta}T \tag{2.70}$$

since it can be shows that $I(1:2;\omega_1)$ with $\omega_1 = X(0)$ is the first term on the right-hand side of (2.70) (Kullback (1959, p. 50)). The value in (2.70) is the same as that in (2.55) using the density given by Yaglom.

To obtain the log-likelihood ratio $\ell(X(t))$ we shall find the limit

$$\ell(X(t)) = ln \frac{g_1(x(0))}{g_2(x(0))} + \mathbf{lim} \sum ln \frac{h_1(x(t+\Delta t)|x(t))}{h_2(x(t+\Delta t)|x(t))} \tag{2.71}$$

where

$$g_i(x(o)) = \frac{1}{\sigma_i(2\pi)^{\frac{1}{2}}} exp\left\{-\frac{x^2(o)}{2\sigma_i^2}\right\}, \quad i = 1,2 \tag{2.72}$$

$$h_i(x(t+\Delta t)|(x(t)) = \frac{1}{\sigma_i(2\pi(1-\rho_i^2))^{\frac{1}{2}}} exp\left\{-\frac{(x(t+\Delta t)-\beta_i x(t))^2}{2\sigma_i^2(1-\rho_i^2)}\right\}, \quad i = 1,2 \tag{2.73}$$

where $\beta_i, \sigma_1^2, \sigma_2^2$ are given in (2.65), (2.66) and we recall (2.67). We thus have for the right-hand side of (2.71) to within terms of order Δt

$$\begin{aligned} &\tfrac{1}{2} ln \frac{\beta}{\alpha} - \frac{\beta x^2(0)}{2} + \frac{\alpha x^2(0)}{2} + \\ &+ \sum \{\tfrac{1}{2} ln \frac{\beta}{\alpha} + \tfrac{1}{2} ln \frac{1-\rho_2^2}{1-\rho_1^2} - \frac{\beta(x(t+\Delta t)-\rho_1 x(t))^2}{2(1-\rho_1^2)} \\ &\qquad + \frac{\alpha(x(t+\Delta t)-\rho_2 x(t))^2}{2(1-\rho_2^2)}\} \end{aligned} \tag{2.74}$$

$$= \tfrac{1}{2}\,ln\,\frac{\beta}{\alpha} + \frac{\alpha-\beta}{2}\,x^2(0) + \sum \{\tfrac{1}{2}\,ln\,\frac{\beta}{\alpha} + \tfrac{1}{2}\,ln\,\frac{(2-\alpha\Delta t)(\alpha\Delta t - \frac{\alpha^2}{2}(\Delta t)^2)}{(2-\beta\Delta t)(\beta\Delta t - \frac{\beta^2}{2}(\Delta t)^2)}$$

$$- \frac{(x(t) + x'(t)\Delta t - x(t) + \beta x(t)\Delta t)^2}{2\Delta t(2-\beta\Delta t)} + \frac{(x(t) + x'(t)\Delta t - x(t) + \alpha x(t)\Delta t)^2}{2\Delta t(2-\beta\Delta t)}$$

$$= \tfrac{1}{2}\,ln\,\frac{\beta}{\alpha} + \frac{\alpha-\beta}{2}\,x^2(0) + \sum \{\tfrac{1}{2}\,ln\,\frac{\beta}{\alpha} + \tfrac{1}{2}\,ln\,\frac{\alpha\Delta t(1-\alpha\Delta t)}{\beta\Delta t(1-\beta\Delta t)}$$

$$- \frac{(\beta x(t) + x'(t))^2\,\Delta t}{4} + \frac{(\alpha x(t) + x'(t))^2\,\Delta t}{4}\}$$

$$= \tfrac{1}{2}\,ln\,\frac{\beta}{\alpha} + \frac{\alpha-\beta}{2}\,x^2(0) + \tfrac{1}{2}(\beta-\alpha)\int_0^T dt - \frac{2\beta}{4}\int_0^T x(t)\,x'(t)\,dt$$

$$+ \frac{2\alpha}{4}\int_0^T x(t)\,x'(t)\,dt + \frac{\alpha^2-\beta^2}{4}\int_0^T x^2(t)\,dt$$

$$= \tfrac{1}{2}\,ln\,\frac{\beta}{\alpha} + \frac{\alpha-\beta}{2}\,x^2(0) + \tfrac{1}{2}(\beta-\alpha)\,T + \frac{\alpha-\beta}{4}\,(x^2(T) - x^2(0)) +$$

$$+ \frac{\alpha^2-\beta^2}{4}\int_0^T x^2(t)\,dt$$

$$= \tfrac{1}{2}\,ln\,\frac{\beta}{\alpha} + \frac{\alpha-\beta}{4}\,(x^2(0) + x^2(T) - 2T) + \frac{\alpha^2-\beta^2}{4}\int_0^T x^2(t)\,dt\,.$$

It is seen that the value of the log-likelihood ratio given by the last expression in (2.74) agrees with the value given in (2.54) and (2.62).

Gaussian Markov Process

As another example of a case in which the means are the same under both hypotheses but the covariance functions are different let us consider the discrete time stationary Gaussian Markov processes with covariance functions

$$H_1 : B_1(t) = \sigma^2 exp(-a|t|),\ a > 0,$$
$$H_2 : B_2(t) = \sigma^2 exp(-b|t|),\ b > 0,\ t = \ldots -1, 0, 1, \ldots . \tag{2.75}$$

In this case the variances are the same under both hypotheses and only the correlations between successive time periods are different under the two hypotheses

$$H_1 : \rho_1 = exp(-a) = \alpha,\ H_2 : \rho_2 = exp(-b) = \beta . \tag{2.76}$$

The corresponding spectral densities are (see Yaglom (1962, p. 57))

$$H_1 : f_1(\lambda) = \frac{\sigma^2(1-\alpha^2)}{2\pi|1-\alpha e^{i\lambda}|^2},\ H_2 : f_2(\lambda) = \frac{\sigma^2(1-\beta^2)}{2\pi|1-\beta e^{i\lambda}|^2} \tag{2.77}$$

where

$$f_n(\lambda) = \frac{1}{2\pi} \sum_{t=-\infty}^{\infty} e^{-i\lambda t} B_n(t),\quad i = (-1)^{1/2},\quad n = 1, 2 \tag{2.78}$$

and

$$B_n(t) = \int_{-\pi}^{\pi} e^{i\lambda t} f_n(\lambda)\, d\lambda,\quad n = 1, 2 . \tag{2.79}$$

The rate of discrimination information is given by (see (2.56))

$$\bar{I} = \lim_{n\to\infty} \frac{1}{n} I(1:2; x(t),\ -n \le t \le n)$$
$$= \frac{1}{2\pi} \int_{-\pi}^{\pi} \left(\frac{f_1(\lambda)}{f_2(\lambda)} - 1 - ln \frac{f_1(\lambda)}{f_2(\lambda)} \right) d\lambda \tag{2.80}$$

$$= \frac{1}{2\pi} \int_{-\pi}^{\pi} \left(\frac{(1-\alpha^2)(1-2\beta\cos\lambda+\beta^2)}{(1-\beta^2)(1-2\alpha\cos\lambda+\alpha^2)} - 1 - ln \frac{(1-\alpha^2)(1-2\beta\cos\lambda+\beta^2)}{(1-\beta^2)(1-2\alpha\cos\lambda+\alpha^2)} \right) d\lambda \,.$$

Using Peirce (1929, no. 304 p. 42, no. 300 p. 41), it is found that

$$\int_{-\pi}^{\pi} \frac{1-2\beta\cos\lambda+\beta^2}{1-2\alpha\cos\lambda+\alpha^2}\, d\lambda = 2 \int_{0}^{\pi} \frac{1-2\beta\cos\lambda+\beta^2}{1-2\alpha\cos\lambda+\alpha^2} \qquad (2.81)$$

$$= 2\left\{ \frac{\beta\pi}{\alpha} + \frac{(1+\beta^2)\,\alpha - (1+\alpha^2)\,\beta}{\alpha} \int_0^{\pi} \frac{d\lambda}{1-2\alpha\cos\lambda+\alpha^2}\, d\lambda \right\}$$

$$= 2\left\{ \frac{\beta\pi}{\alpha} + \frac{(\alpha-\beta)(1-\alpha\beta)}{\alpha} \frac{2}{1-\alpha^2} \frac{\pi}{2} \right\} = \frac{1-2\alpha\beta+\beta^2}{1-\alpha^2}\, 2\pi \,,$$

and from Peirce (1929, no. 523 p. 65)

$$\int_0^{\pi} ln\,(1-2\beta\cos\lambda+\beta^2)\, d\lambda = 0 = \int_0^{\pi} ln\,(1-2\alpha\cos\lambda+\alpha^2)\, d\lambda \,. \qquad (2.82)$$

Now using (2.81) and (2.82) in (2.80) it is found that

$$\bar{I} = \frac{1-2\alpha\beta+\beta^2}{1-\beta^2} - 1 - ln \frac{1-\alpha^2}{1-\beta^2} \qquad (2.83)$$

$$= \frac{2\beta(\beta-\alpha)}{1-\beta^2} + ln \frac{1-\beta^2}{1-\alpha^2} \,.$$

To use the relation for the discrimination information values and densities corresponding to the Markov property we see from (2.65) that

$$I(1:2\,;x(t+1)\,|\,x(t)) = \tfrac{1}{2}\, ln \frac{1-\rho_2^2}{1-\rho_1^2} - \tfrac{1}{2} + \frac{1-\rho_1^2}{2(1-\rho_2^2)} + \frac{(\rho_2-\rho_1)^2}{2(1-\rho_2^2)} \qquad (2.84)$$

since the variances are the same under the two hypotheses. Thus

$$I(1:2;x(t),\ -n \le t \le n) = 2nI(1:2;x(t+1)\,|\,x(t)) \tag{2.85}$$

since there are $2n$ transitions and $I(1:2;\mathrm{x}(\mathrm{t})) = 0$. Accordingly

$$\lim_{n\to\infty} \frac{1}{n} I(1:2;x(t),\ -n \le t \le n) = \frac{2\rho_2(\rho_2 - \rho_1)}{1-\rho_2^2} + ln\,\frac{1-\rho_2^2}{1-\rho_1^2} \tag{2.86}$$

$$= \frac{2\beta\,(\beta - \alpha)}{1-\beta^2} + ln\,\frac{1-\beta^2}{1-\alpha^2}$$

that is, the value found in (2.83).

Note that for $\beta = 0$, $f_2(\lambda) = \sigma^2/2\pi$, $B_2(0) = \sigma^2$, $B_2(t) = 0$, and in this case H_2 implies noise, that is, a process with spectral density a constant and $B_2(t) = 0$ for $t \neq 0$. The discrimination information values then correspond to a test of Markovity of order one against independence.

It is of interest to examine the results in (2.56) and (2.80) with conditions on the spectral densities under which the rate of discrimination information is given as in (2.56) and (2.80). It has been shown (Pinsker (1964, p. 199)) that in the discrete-parameter case, (2.80) holds if

$$\frac{f_1(\lambda)}{f_2(\lambda)} < C \tag{2.87}$$

where C is a constant independent of λ. This condition is satisfied for the spectral densities in (2.77) since $\alpha \neq 1$ and hence $1 - 2\alpha\cos\lambda + \alpha^2 \neq 0$. In the continuous parameter case, (2.56) holds if

$$\lim_{\lambda\to\infty} \frac{f_1(\lambda)}{f_2(\lambda)} = 1\,, \tag{2.88}$$

and we note that the spectral densities in (2.53) satisfy (2.88). On the other hand if

$$\lim_{\lambda \to \infty} \frac{f_1(\lambda)}{f_2(\lambda)} \neq 1 , \tag{2.89}$$

then the likelihood ratio is infinite for every $T > 0$ and the discrimination information is infinite for every T. Thus if we consider the continuous time stationary Gaussian Markov processes with covariance functions

$$H_1 : B_1(t) = \sigma^2 exp(-a|t|),\ a > 0 \tag{2.90}$$

$$H_2 : B_2(t) = \sigma^2 exp(-b|t|),\ b > 0,\ -\infty < t < \infty$$

and with spectral densities

$$f_n(\lambda) = \frac{1}{2\pi} \int_{-\infty}^{\infty} e^{-i\lambda t} B_n(t)\, dt,\ n = 1, 2 \tag{2.91}$$

$$f_1(\lambda) = \frac{\sigma^2 a}{\pi(\lambda^2 + a^2)},\ f_2(\lambda) = \frac{\sigma^2 b}{\pi(\lambda^2 + b^2)}$$

that is, only the correlations are different under the two hypotheses, we see that

$$\lim_{\lambda \to \infty} \frac{f_1(\lambda)}{f_2(\lambda)} = \lim_{\lambda \to \infty} \frac{a(\lambda^2 + b^2)}{b(\lambda^2 + a^2)} = \frac{a}{b} \neq 1 . \tag{2.92}$$

In this case we find that the integral in (2.56) does not yield a finite value nor does the procedure similar to that in (2.65) - (2.70) yield a finite value, nor does the procedure similar to that in (2.71) - (2.74) yield a finite value.

It should be noted that the discrimination information in (2.5) and the limit in (2.7) are not restricted only to the case of stationary Markov processes as in the examples discussed. Thus if $X(t)$ is a stochastic process with time t discrete and if we denote $X(m) = \omega_m$, then (2.5) corresponds to the discrimination information when the hypothesis H_i, $i = 1, 2$, specifies the joint density $f_i(\omega_1, \omega_2, \ldots, \omega_n)$, $i = 1, 2$ for the random variables $\omega_1 = X(1), \omega_2 = X(2), \ldots, \omega_n = X(n)$ respectively. When the time t is continuous, $0 \le t \le T$, and $0 \le t_1 \le \ldots \le t_n = T$, and if we denote $X(t_m) = \omega_m$, then (2.5) corresponds to the discrimination information when the hypothesis H_i, $i = 1, 2$, specifies the joint density $f_i(\omega_1, \omega_2, \ldots, \omega_n)$, $i = 1, 2$ for the random variables $\omega_1 = X(t_1), \omega_2 = X(t_2), \ldots, \omega_n = X(t_n)$ respectively. The limit in (2.7) corresponds to a limit taken over a sequence of finer partitions $0 \le t_1 \le \ldots \le t_n = T$, $max_i(t_{i+1} - t_i) \to 0$ and we shall write, as before

$$\lim_{n \to \infty} I(1:2;\beta_n) = I(1:2;\beta) = I(1:2;X(t),\ 0 \le t \le T). \tag{2.93}$$

V: LINEAR DISCRIMINANT FUNCTIONAL

1. Representation of a Random Vector

In this chapter we shall consider examples illustrating further applications of the concepts already developed to a number of stochastic processes. In particular, in the examples we shall consider, the use of a linear discriminant functional is of great utility. Before taking up the stochastic processes we believe it desirable to examine first the representation of a random vector and information theoretic techniques as applied in this case. We shall follow procedures which lend themselves to generalization to the case of stochastic processes.

It is known that there exists a nonsingular matrix $\mathbf{A}$ such that

$$\mathbf{A}\Sigma\mathbf{A}' = \Lambda \ , \ \mathbf{AA}' = \mathbf{I} \tag{1.1}$$

where $\boldsymbol{\Sigma}$ is the covariance matrix of $\mathbf{x}' = (x(1), x(2), \ldots, x(n))$, $\boldsymbol{\Lambda}$ is a diagonal matrix with diagonal elements the characteristic roots (eigen values) of $\boldsymbol{\Sigma}$ and $\mathbf{I}$ is the nxn identity matrix, that is, $\mathbf{A}$ is an orthogonal matrix (e.g., in Kullback (1959, p. 195) set $\boldsymbol{\Sigma}_2 = \mathbf{I}$, $\boldsymbol{\Sigma}_1 = \boldsymbol{\Sigma}$). We can also write

$$\Sigma = \mathbf{A}'\Lambda\mathbf{A} . \tag{1.2}$$

Let us write

$$\mathbf{A}' = (\boldsymbol{\alpha}_1, \boldsymbol{\alpha}_2, \ldots, \boldsymbol{\alpha}_n), \quad \mathbf{A} = \begin{pmatrix} \alpha_1' \\ \cdot \\ \cdot \\ \cdot \\ \alpha_n' \end{pmatrix}, \quad \boldsymbol{\alpha}_j' = (\alpha_j(1), \alpha_j(2), \ldots \alpha_j(n))$$

so that (1.2) may be written as

$$\boldsymbol{\Sigma} = (\boldsymbol{\alpha}_1, \boldsymbol{\alpha}_2, \ldots, \boldsymbol{\alpha}_n) \begin{pmatrix} \lambda_1 & 0 & \ldots 0 \\ 0 & \lambda_2 & \ldots 0 \\ \cdot & \cdot & \cdot \\ 0 & \ldots & \lambda_n \end{pmatrix} \begin{pmatrix} \boldsymbol{\alpha}_1' \\ \cdot \\ \cdot \\ \boldsymbol{\alpha}_n' \end{pmatrix} = \sum_{i=1}^{n} \lambda_i \boldsymbol{\alpha}_i \boldsymbol{\alpha}_i' . \tag{1.3}$$

If we write $\boldsymbol{\Sigma} = (\sigma(s,t))$, $s, t = 1, 2, \ldots, n$, then from (1.3) it is seen that

$$\sigma(s,t) = \sum_{i=1}^{n} \lambda_i \alpha_i(s) \alpha_i(t), \quad s, t = 1, 2, \ldots, n. \tag{1.4}$$

Let $\mathbf{y}$ be a random vector with independent components of respective variances λ_i, then the linear transformation

$$\mathbf{x} = \mathbf{A}'\mathbf{y} \tag{1.5}$$

is such that (1.2) is satisfied. We note that (1.5) may also be written as

$$\mathbf{x} = (\boldsymbol{\alpha}_1, \boldsymbol{\alpha}_2, \ldots, \boldsymbol{\alpha}_n)\,\mathbf{y} = \boldsymbol{\alpha}_1 y_1 + \boldsymbol{\alpha}_2 y_2 + \ldots + \boldsymbol{\alpha}_n y_n \tag{1.6}$$

or

$$x(t) = y_1 \alpha_1(t) + y_2 \alpha_2(t) + \ldots + y_n \alpha_n(t), \quad t = 1, 2, \ldots, n. \tag{1.7}$$

The characteristic roots (eigenvalues) λ_i and characteristic vectors (eigenvectors) $\boldsymbol{\alpha}_i$ satisfy

$$\Sigma \boldsymbol{\alpha}_i = \lambda_i \boldsymbol{\alpha}_i \tag{1.8}$$

or

$$\sum_{t=1}^{n} \sigma(s,t)\, \alpha_i(t) = \lambda_i \alpha_i(s),\ s = 1, 2, \ldots, n. \tag{1.9}$$

The conditions $\mathbf{A}\mathbf{A}' = \mathbf{A}'\mathbf{A} = \mathbf{I}$ imply that

$$\boldsymbol{\alpha}_i' \boldsymbol{\alpha}_i = 1,\ \boldsymbol{\alpha}_i' \boldsymbol{\alpha}_j = 0,\ \sum_{i=1}^{n} \boldsymbol{\alpha}_i \boldsymbol{\alpha}_i' = \mathbf{I} \tag{1.10}$$

or

$$\sum_{t=1}^{n} \alpha_i(t)\, \alpha_i(t) = 1,\ \sum_{t=1}^{n} \alpha_i(t)\, \alpha_j(t) = 0,\ \sum_{i=1}^{n} \alpha_i(s)\, \alpha_i(t) = \delta_{st} = \begin{cases} 1 & s = t \\ 0 & s \neq t \end{cases}. \tag{1.11}$$

Let us now define the random variable z as a linear combination of the components of $\mathbf{x}$, that is, $z = \mathbf{b}'\mathbf{x} = \sum_{t=1}^{n} b(t)\, x(t)$. It follows that

$$\mu_z = E(z) = E(\mathbf{b}'\mathbf{x}) = \mathbf{b}'\boldsymbol{\mu}_x = \sum_{t=1}^{n} b(t)\, \mu(t), \tag{1.12}$$

$$\sigma_z^2 = E(z-\mu_z)^2 = E(\mathbf{b}'(\mathbf{x}-\boldsymbol{\mu}_x))^2 = E(\mathbf{b}'(\mathbf{x}-\boldsymbol{\mu}_x)(\mathbf{x}-\boldsymbol{\mu}_x)'\mathbf{b}) = \mathbf{b}'\Sigma\, \mathbf{b}. \tag{1.13}$$

If we consider $n \to \infty$, and if $\mathbf{b}'\Sigma\, \mathbf{b} \to 0$ then $E(\mathbf{b}'\mathbf{x} - \mathbf{b}'\boldsymbol{\mu}_{\mathbf{x}})^2 \to 0$, that is, l.i.m. $(\mathbf{b'x} - \mathbf{b'\mu_x}) = 0 \Rightarrow \sum_{t=1}^{n} b(t)x(t) \to \sum_{t=1}^{n} b(t)\mu(t)$ in probability.

In particular if we take $b(t) = 1/n$, $\mu(t) = \mu$, then

$$\lim_{n\to\infty} \frac{1}{n} \sum_{t=1}^{n} x(t) \to \mu, \quad \text{in probability} \tag{1.14}$$

provided that

$$\mathbf{b'}\,\Sigma\,\mathbf{b} = \lim_{n\to\infty} \frac{1}{n^2} \sum_{s=1}^{n} \sum_{t=1}^{n} \sigma(s,t) = 0 . \tag{1.15}$$

If the spectral representation of $\boldsymbol{\Sigma}$ given by (1.3) or (1.4) is used, then (1.15) may be written as

$$\lim_{n\to\infty} \frac{1}{n^2} \sum_{s=1}^{n} \sum_{t=1}^{n} \sum_{i=1}^{n} \lambda_i\, \alpha_i(s)\, \alpha_i(t) = 0 . \tag{1.16}$$

In particular if $|\lambda_i| \le \lambda < \infty$, then using (1.11)

$$\frac{1}{n^2} \sum_{s=1}^{n} \sum_{t=1}^{n} \sum_{i=1}^{n} \lambda_i\, \alpha_i(s)\alpha_i(t) \le \frac{\lambda}{n^2} \sum_{s=1}^{n} \sum_{t=1}^{n} \delta_{st} = \frac{\lambda}{n} \tag{1.17}$$

and condition (1.16) is satisfied, that is, the ergodic property (1.14) holds.

For the stationary (or wide sense stationary) case, in which $\sigma(s,t) = \sigma(s-t) = \sigma(t-s)$, (1.15) may be written as

$$\frac{1}{n^2} \sum_{s} \sum_{t} \sigma(s,t) = \frac{1}{n^2} \sum_{r=-n}^{n} (n - |r|)\sigma(r) = \frac{1}{n} \sum_{r=-n}^{n} \left(1 - \frac{|r|}{n}\right)\sigma(r) \tag{1.18}$$

and thus the ergodic property (1.14) holds if

$$\lim_{n\to\infty} \frac{1}{n} \sum_{r=-n}^{n} \left(1 - \frac{|r|}{n}\right) \sigma(r) \to 0 . \tag{1.19}$$

Using the Herglotz theorem (Loève (1955) that

$$\sigma(r) = \int_{-\pi}^{\pi} e^{ir\lambda}\, dF(\lambda), \tag{1.20}$$

where F is a nondecreasing bounded function, the spectral distribution function of $x(t)$, (1.19) yields

$$\lim_{n\to\infty} \frac{1}{n} \sum_{r=-n}^{n} \left(1 - \frac{|r|}{n}\right) \sigma(r) = \lim_{n\to\infty} \frac{1}{n} \sum_{r=-n}^{n} \left(1 - \frac{|r|}{n}\right) \int_{-\pi}^{\pi} e^{ir\lambda}\, dF(\lambda)$$

$$= \lim_{n\to\infty} \int_{-\pi}^{\pi} dF(\lambda) \left\{ \frac{1}{n} \sum_{r=-n}^{n} \left(1 - \frac{|r|}{n}\right) e^{ir\lambda} \right\}$$

$$= \lim_{n\to\infty} \int_{-\pi}^{\pi} \frac{\sin^2 \frac{n\lambda}{2}}{n^2 \sin^2 \frac{\lambda}{2}}\, dF(\lambda) = F(+0) - F(-0) . \tag{1.21}$$

Thus the ergodic property holds if $F(+0) - F(-0) = 0$, that is, there is no positive spectral mass at the origin.

Another approach to these results is to write

$$\frac{1}{n^2} \sum_s \sum_t \sigma(s,t) = \frac{1}{n^2} \sum_s \sum_t \int_{-\pi}^{\pi} e^{i\lambda(s-t)}\, dF(\lambda)$$

$$= \frac{1}{n^2} \int_{-\pi}^{\pi} dF(\lambda) \left\{ \sum_0^{n-1} \exp is\lambda \right\} \left\{ \sum_0^{n-1} \exp -it\lambda \right\}$$

$$= \frac{1}{n^2} \int_{-\pi}^{\pi} dF(\lambda) \; \frac{1-e^{in\lambda}}{1-e^{i\lambda}} \cdot \frac{1-e^{-in\lambda}}{1-e^{-i\lambda}} = \frac{1}{n^2} \int_{-\pi}^{\pi} dF(\lambda) \; \frac{1-\cos n\lambda}{1-\cos\lambda}$$

$$= \int_{-\pi}^{\pi} \frac{\sin^2 \frac{n\lambda}{2}}{n^2 \sin^2 \frac{\lambda}{2}} \, dF(\lambda) \,. \tag{1.22}$$

Since $\frac{1}{n} \sum_{r=-n}^{n} (1 - \frac{|r|}{n}) = 1$, we also note that $\frac{1}{n} \sum_{r=-n}^{n} (1 - \frac{|r|}{n}) \, e^{i\lambda r}$

$$= (\frac{1}{n} \sum_{0}^{n-1} \exp is\lambda) \; (\frac{1}{n} \sum_{0}^{n-1} \exp -it\lambda)$$

is the characteristic function of $z = x - y$ where x and y are identically distributed random variables such that Prob $\{x = r\} = 1/n$, $r = 0, 1, \ldots, n-1$. The inversion formula for the characteristic function yields

$$\frac{1}{n} (1 - \frac{|r|}{n}) = \frac{1}{2\pi} \int_{-\pi}^{\pi} \frac{e^{-i\lambda r} \sin^2 \frac{n\lambda}{2}}{n^2 \sin^2 \frac{\lambda}{2}} \, d\lambda \,, \; r = 0, \pm 1, \ldots, \pm n \tag{1.23}$$

from which it is seen that

$$\lim_{n \to \infty} \frac{1}{2\pi} \int_{-\pi}^{\pi} \frac{e^{-i\lambda r} \sin^2 \frac{n\lambda}{2}}{n \sin^2 \frac{\lambda}{2}} \, d\lambda = 1 \,, \quad \text{for any fixed } r \,. \tag{1.24}$$

For the linear discriminant function $z = \mathbf{b}'\mathbf{x}$ it was seen that

$$\sigma_z^2 = \mathbf{b}'\boldsymbol{\Sigma}\,\mathbf{b} = \mathbf{b}'\Big(\sum_{i=1}^{n} \lambda_i \boldsymbol{\alpha}_i \boldsymbol{\alpha}_i'\Big)\mathbf{b} = \sum_{i=1}^{n} \lambda_i \mathbf{b}'\boldsymbol{\alpha}_i \boldsymbol{\alpha}_i'\mathbf{b} = \sum_{i=1}^{n} \lambda_i c_i^2 \tag{1.25}$$

where $c_i = \mathbf{b}'\boldsymbol{\alpha}_i = \boldsymbol{\alpha}_i'\mathbf{b} = \sum_{t=1}^{n} b(t)\,\alpha_i(t)$. Multiplying by $\boldsymbol{\alpha}_i$ and summing, one obtains (see (1.10))

$$\sum_{i=1}^{n} c_i \boldsymbol{\alpha}_i = \sum_{i=1}^{n} \boldsymbol{\alpha}_i \boldsymbol{\alpha}_i' \mathbf{b} = \mathbf{b} \tag{1.26}$$

that is,

$$b(t) = \sum_{i=1}^{n} c_i\, \alpha_i(t),\ c_i = \sum_{t=1}^{n} b(t)\, \alpha_i(t) \tag{1.27}$$

or a Fourier expansion of $b(t)$.

If we now assume that the *y's* in (1.5) or (1.6) are normally distributed with $E_1(\mathbf{y}) = \mathbf{m}_1$, $E_2(\mathbf{y}) = \mathbf{m}_2$, and $\mathrm{cov}_1(\mathbf{y}) = \mathrm{cov}_2(\mathbf{y}) = \boldsymbol{\Lambda}$, and write $\mathbf{m} = \mathbf{m}_1 - \mathbf{m}_2$, then $\mathbf{x}$ defined in (1.5) or (1.6) is normally distributed with

$$E_1(\mathbf{x}) = \mathbf{A}'\mathbf{m}_1 = \boldsymbol{\mu}_1,\ E_2(\mathbf{x}) = \mathbf{A}'\mathbf{m}_2 = \boldsymbol{\mu}_2,\ E_1(\mathbf{x}) - E_2(\mathbf{x}) = \mathbf{A}'\mathbf{m} = \boldsymbol{\mu}, \tag{1.28}$$

$$\boldsymbol{\Sigma} = \mathbf{A}'\boldsymbol{\Lambda}\,\mathbf{A} = \sum_{i=1}^{n} \lambda_i \boldsymbol{\alpha}_i \boldsymbol{\alpha}_i', \tag{1.29}$$

$$\boldsymbol{\mu} = \mathbf{A}'\mathbf{m} = (\boldsymbol{\alpha}_1, \boldsymbol{\alpha}_2, \ldots, \boldsymbol{\alpha}_n)\,\mathbf{m} = \boldsymbol{\alpha}_1 m_1 + \boldsymbol{\alpha}_2 m_2 + \ldots + \boldsymbol{\alpha}_n m_n, \tag{1.30}$$

that is

$$\sigma(s,t)=\sum_{i=1}^{n}\lambda_i\,\alpha_i(s)\,\alpha_i(t),\ \mu(t)=m_1\alpha_1(t)+m_2\alpha_2(t)+\ldots+m_n\alpha_n(t), \tag{1.31}$$

and $z=\mathbf{b}'\mathbf{x}$ is also normally distributed with

$$E_1(z)=\mathbf{b}'\mathbf{A}'\mathbf{m}_1\,,\ E_2(z)=\mathbf{b}'\mathbf{A}'\mathbf{m}_2,\ E_1(z)-E_2(z)=\mathbf{b}'\mathbf{A}'\mathbf{m} \tag{1.32}$$

$$\mathrm{Var}_1(z)=\mathrm{Var}_2(z)=\mathbf{b}'\boldsymbol{\Sigma}\,\mathbf{b}=\sum_{i=1}^{n}\lambda_i c_i^{\,2}\,. \tag{1.33}$$

For the respective normal distributions it follows that

$$2I(1:2\,;\mathbf{y})=\mathbf{m}'\boldsymbol{\Lambda}^{-1}\mathbf{m}=\sum_{i=1}^{n}\frac{m_i^2}{\lambda_i}\,, \tag{1.34}$$

$$2I(1:2\,;\mathbf{x})=\boldsymbol{\mu}'\boldsymbol{\Sigma}^{-1}\boldsymbol{\mu}=\mathbf{m}'\mathbf{A}(\mathbf{A}'\boldsymbol{\Lambda}\,\mathbf{A})^{-1}\mathbf{A}'\mathbf{m}=\mathbf{m}'\boldsymbol{\Lambda}^{-1}\mathbf{m}\,, \tag{1.35}$$

$$2I(1:2\,;z)=\frac{\mathbf{b}'\boldsymbol{\mu}\boldsymbol{\mu}'\mathbf{b}}{\mathbf{b}'\boldsymbol{\Sigma}\,\mathbf{b}}=\frac{\left(\sum_{i=1}^{n}c_i\,\boldsymbol{\alpha}_i'\boldsymbol{\mu}\right)^2}{\sum_{i=1}^{n}\lambda_i c_i^{\,2}}=\frac{\left(\sum_{i=1}^{n}c_i d_i\right)^2}{\sum_{i=1}^{n}\lambda_i c_i^{\,2}} \tag{1.36}$$

where $d_i=\boldsymbol{\alpha}_i'\mu=\sum_{t=1}^{n}\alpha_i(t)\,\mu(t)=\sum_{t=1}^{n}\sum_{j=1}^{n}\alpha_i(t)\,m_j\,\alpha_j(t)=m_i\,.$

Writing $\sum_{i=1}^{n}c_i\,d_i=\sum_{t=1}^{n}(c_i\,(\lambda_i)^{1/2})\,(d_i\,/\,(\lambda_i)^{1/2})$ and using the Schwarz inequality

$$\left(\sum_{i=1}^{n} c_i d_i\right)^2 \le \left(\sum_{i=1}^{n} \lambda_i c_i^2\right)\left(\sum_{i=1}^{n} d_i^2/\lambda_i\right) \tag{1.37}$$

with equality if and only if $c_i\,(\lambda_i)^{1/2} = d_i\,/\,(\lambda_i)^{1/2}$, or

$$d_i = c_i \lambda_i = \boldsymbol{\alpha}_i' \boldsymbol{\mu} \tag{1.38}$$

$$\sum_{i=1}^{n} \lambda_i \boldsymbol{\alpha}_i c_i = \sum_{i=1}^{n} \boldsymbol{\alpha}_i \boldsymbol{\alpha}_i' \boldsymbol{\mu} \tag{1.39}$$

$$\sum_{\mathbf{i}=1}^{\mathbf{n}} \lambda_i \boldsymbol{\alpha}_{\mathbf{i}} \boldsymbol{\alpha}_{\mathbf{i}}' \mathbf{b} = \boldsymbol{\mu} = \boldsymbol{\Sigma}\mathbf{b} \tag{1.40}$$

it is seen that (1.36) is a maximum when (1.38) – (1.40) are satisfied and that

$$\text{Max } 2I(1:2\,;\mathrm{z}) = \max \frac{\left(\sum_{i=1}^{n} c_i d_i\right)^2}{\sum_{i=1}^{n} \lambda_i c_i^2} = \sum_{i=1}^{n} \lambda_i c_i^2 = \sum_{i=1}^{n} \frac{d_i^2}{\lambda_i}$$

$$= \sum_{i=1}^{n} \frac{\boldsymbol{\mu}' \boldsymbol{\alpha}_{\mathrm{i}} \boldsymbol{\alpha}_{\mathrm{i}}' \boldsymbol{\mu}}{\lambda_i} = \boldsymbol{\mu}' \left(\sum_{i=1}^{n} \frac{\boldsymbol{\alpha}_{\mathrm{i}} \boldsymbol{\alpha}_{\mathrm{i}}'}{\lambda_i}\right) \boldsymbol{\mu}$$

$$= \boldsymbol{\mu}' \boldsymbol{\Sigma}^{-1} \boldsymbol{\mu} = 2I(1{:}\,2;\mathbf{x}) = \sum_{i=1}^{n} \frac{m_i^2}{\lambda_i} = 2\,I(1\,;2\,;\mathbf{y})\,, \tag{1.41}$$

and the corresponding value of z is given by

$$z = \mathbf{b}'\mathbf{x} = \sum_{i=1}^{n} c_i \boldsymbol{\alpha}_i' \mathbf{x} = \sum_{i=1}^{n} c_i \boldsymbol{\alpha}_i' \sum_{j=1}^{n} \boldsymbol{\alpha}_j y_j = \sum_{i=1}^{n} c_i y_i = \sum_{i=1}^{n} \frac{d_i y_i}{\lambda_i} . \tag{1.42}$$

We note that the moment generating function of $\mathbf{x}$ under H_2 is given by

$$M_2(\mathbf{t}) = E_2(exp\, \mathbf{t}'\mathbf{x}) = exp\, (\mathbf{t}'\boldsymbol{\mu}_2 + \frac{1}{2}\mathbf{t}'\boldsymbol{\Sigma}\mathbf{t}) . \tag{1.43}$$

Accordingly for the minimum discrimination information distribution the condition

$$\boldsymbol{\theta}^* = \frac{d}{d\mathbf{t}} \, ln\, M_2(\mathbf{t}) \tag{1.44}$$

becomes

$$\boldsymbol{\theta}^* = \boldsymbol{\mu}_2 + \boldsymbol{\Sigma}\mathbf{t} . \tag{1.45}$$

Hence for $\boldsymbol{\theta}^* = \boldsymbol{\mu}_1$ and $\mathbf{t} = \mathbf{b}$ (1.45) is the same condition as (1.40). The log-likelihood ratio is

$$ln\, \frac{f^*}{f_2} = \mathbf{t}'\mathbf{x} - ln\, M_2(\mathbf{t})$$

$$= \boldsymbol{\mu}'\boldsymbol{\Sigma}^{-1}\mathbf{x} - \boldsymbol{\mu}'\boldsymbol{\Sigma}^{-1}\boldsymbol{\mu}_2 - \frac{1}{2}\boldsymbol{\mu}'\boldsymbol{\Sigma}^{-1}\boldsymbol{\mu}$$

$$= \boldsymbol{\mu}'\boldsymbol{\Sigma}^{-1}(\mathbf{x} - \boldsymbol{\mu}_2) - \frac{1}{2}\boldsymbol{\mu}'\boldsymbol{\Sigma}^{-1}\boldsymbol{\mu}$$

$$= \sum_{i=1}^{n} \frac{d_i (y_i - m_{2i})}{\lambda_i} - \frac{1}{2} \sum_{i=1}^{n} \frac{m_i^2}{\lambda_i} . \tag{1.46}$$

We shall encounter appropriate analogues of the various results in this section in the examples discussing certain stochastic processes.

The Dirac Delta-function

In the following examples we shall have occasion to use the Dirac delta-function which we shall now describe. See for example Rényi (1962). The Dirac delta-function $\delta(x)$ is such that $\delta(x) = 0$ except for $x = 0$ where it tends to ∞ in such a way that $\int\delta(x)\,dx = 1$, for example, with $f_n(x) =$

$(n/2\pi)^{1/2} \, exp\,(-nx^2/2)$, $\lim_{n\to\infty} \int_{-\infty}^{\infty} f_n(x)\,dx = \int \delta(x)\,dx$. With $i = (-1)^{1/2}$,

$\int_{-\infty}^{\infty} \exp\,(i\,(t-s)\,\lambda)\,d\lambda = \int_{-\infty}^{\infty} \exp\,(-i\,(t-s)\,\lambda)\,d\lambda = 2\pi\delta\,(t-s)$ and

differentiating $\int_{-\infty}^{\infty} (i\lambda)^k \, exp\,(i(t-s)\lambda)\,d\lambda = 2\pi\delta^{(k)}(t-s)$, where δ is the

Dirac delta – function $\lim_{T\to\infty} (1/2T) \int_{-T}^{T} exp\,(i\,(t-s)\,\lambda)\,d\lambda = \delta_{t,s} = 0$
$t \neq s$ and $= 1$, $t=s$.

The characteristic function of the step distribution function $\varepsilon(x-a)=0$,

$x < a$ and $\varepsilon(x-a) = 1, x \geq a$ is $E\,(exp\,ix\lambda) = \int exp\,(i\lambda x)\,d\varepsilon\,(x-a) = exp\,(i\lambda a)$.

Formally $d\varepsilon(x-a)/dx = (1/2\pi) \int_{-\infty}^{\infty} exp\,(-i\lambda\,x + i\lambda a)\,d\lambda =$

$(1/2\pi) \int_{-\infty}^{\infty} exp\,(-i\lambda\,(x-a))\,d\lambda = \delta(x-a)$, so that $\int f(x)\,\delta\,(x-a)\,dx =$

$\int f(x)\,d(\varepsilon(x-a)) = f(a)$.

Note that if the range of λ is $(-W, W)$ then $\delta(\tau) =$

$$(1/2W)\int_{-W}^{W} \exp(i\tau\lambda)\,d\lambda = (\exp(i\tau W) - \exp(-i\tau W))/2Wi\tau = (\sin\tau W)/W\tau .$$

2. Examples

Example 2.1 Consider the representation of the stochastic process $X(t)$ (see Borpujari and Kullback (1983))

$$X(t) = \sum_{1}^{\infty} y_i \alpha_i(t) \tag{2.1.1}$$

where the y_i are independent normal random variables with $\mathrm{Var}(y_i) = \lambda_i$ and $\{a_i(t)\}$ is a complete set of orthonormal functions of t. The covariance function of $X(t)$ is given by

$$\sigma(s,t) = \mathrm{cov}(X(s), X(t)) = \sum_{1}^{\infty} \lambda_i \alpha_i(s)\alpha_i(t). \tag{2.1.2}$$

If we assume that $E(y_i) = m_i$, then

$$E(X(t)) = m(t) = \sum_{1}^{\infty} m_i \alpha_i(t). \tag{2.1.3}$$

The moment generating functional (m.g.f.) of $X(t)$ is

$$M = E(\exp \int u(t) X(t)\, dt) \tag{2.1.4}$$

where $u(t)$ is the m.g.f. parameter and the integral in (2.1.4) is a linear functional of $X(t)$ defined as a mean square limit. Using the representation (2.1.1) in (2.1.4)

$$M = E\left(exp \int u(t)\left(\sum_1^\infty y_i \alpha_i(t)\right) dt = E\left(exp \sum_1^\infty y_i u_i\right)\right. \tag{2.1.5}$$

where

$$u_i = \int u(t)\, \alpha_i(t)\, dt\ ,\quad u(t) = \sum_1^\infty u_i\, \alpha_i(t)\,. \tag{2.1.6}$$

Since the y's are $N(m_i, \lambda_i)$, (2.1.5) yields

$$M = exp\left\{\sum_1^\infty \left(m_i u_i + \frac{u_i^2 \lambda_i}{2}\right)\right\} \tag{2.1.7}$$

or

$$ln\, M = \sum_1^\infty m_i u_i + \sum_1^\infty \frac{u_i^2 \lambda_i}{2}\ . \tag{2.1.8}$$

Using (2.1.6) , (2.1.3) , and (2.1.2)

$$\sum_1^\infty m_i u_i = \int \left(\sum_i m_i \alpha_i(t)\right) u(t)\, dt = \int m(t)\, u(t)\, dt \tag{2.1.9}$$

$$\sum_1^\infty \frac{\lambda_i u_i^2}{2} = (1/2) \int\int \left(\sum_1^\infty \lambda_i \alpha_i(s)\alpha_i(t)\right) u(s)\, u(t)\, ds\, dt$$

$$= (1/2) \int\int \sigma(s,t)\, u(s)\, u(t)\, ds\, dt . \qquad (2.1.10)$$

Hence for the distribution with minimum discrimination information against that just considered and with expected value of $X(t)$ given by $\theta(t)$, we have

$$f^*(X(t)) = \frac{exp\left\{\int u(t) X(t)\, dt\right\} f(X(t))}{exp\left\{\int m(t)\, u(t)\, dt + (1/2)\int\int \sigma(s,t)\, u(s) u(t)\, ds\, dt\right\}} \qquad (2.1.11)$$

$$I(f^* : f) = \int \theta(t)\, u(t)\, dt - \int m(t)\, u(t)\, dt - (1/2)\int\int \sigma(s,t)\, u(s)\, u(t)\, ds\, dt \qquad (2.1.12)$$

where the f* distribution is exponential. It is to be noted that

$$\theta(t) = \int X(t) f^*(X(t))\, dP = E^*(X(t))$$

$$= \int \frac{X(t)\, exp\left\{\int u(t) X(t)\, dt\right\} f(X(t))\, dP}{M}$$

$$= \frac{E(X(t)\, exp \int u(t) X(t)\, dt)}{E(exp \int u(t) X(t)\, dt)} \qquad (2.1.13)$$

$$\int \theta(t)\, u(t)\, dt = \frac{E(\int u(t) X(t) \left\{exp \int u(t) X(t)\, dt\right\} dt)}{E(exp \int u(t) X(t)\, dt)} . \qquad (2.1.14)$$

Let us define the random variable Z

$$Z = \int u(t) X(t)\, dt, \quad \sigma_z^2 = \iint \sigma(s,t) u(s) u(t)\, ds\, dt \tag{2.1.15}$$

hence (2.1.14) becomes

$$\int \theta(t) u(t)\, dt = \frac{E(Ze^Z)}{E(e^Z)} . \tag{2.1.16}$$

Since Z is a linear functional of the normal process $X(t)$, Z is normally distributed and

$$E(e^{\tau Z}) = exp\left\{ \tau \int m(t) u(t)\, dt + \frac{\tau^2}{2} \iint \sigma(s,t) u(s) u(t)\, ds\, dt \right\} \tag{2.1.17}$$

$$E(Ze^{\tau Z}) = E(e^{\tau Z}) \left\{ \int m(t) u(t)\, dt + \tau \iint \sigma(s,t) u(s) u(t)\, ds\, dt \right\} \tag{2.1.18}$$

hence for $\tau = 1$

$$E(Z e^Z)/E(e^Z) = \int m(t) u(t)\, dt + \iint \sigma(s,t) u(s) u(t)\, ds\, dt, \tag{2.1.19}$$

and (2.1.16) becomes

$$\int \theta(t) u(t)\, dt = \int m(t) u(t)\, dt + \iint \sigma(s,t) u(s) u(t)\, ds\, dt \tag{2.1.20}$$

and (2.1.12) yields

$$I(f^*:f) = (1/2)\int\int \sigma(s,t)\,u(s)\,u(t)\,ds\,dt. \tag{2.1.21}$$

We may also write (2.1.20) as

$$\int [(\theta(t) - m(t)) - \int \sigma(s,t)\,u(s)\,ds\,]\,u(t)\,dt = 0 \tag{2.1.22}$$

and since $u(t)$ is orthogonal only to a function $\equiv 0$,

$$\theta(t) - m(t) = \int \sigma(s,t)\,u(s)\,ds \tag{2.1.23}$$

and

$$I(f^*:f) = (1/2)\int (\theta(t) - m(t))\,u(t)\,dt. \tag{2.1.24}$$

Let us deonte the inverse function of $\sigma(s,t)$ by $\bar{\sigma}(t,v)$ where $\bar{\sigma}(t,v)$ is such that

$$\int \sigma(s,t)\,\bar{\sigma}(t,v)\,dt = \delta(s-v) \tag{2.1.25}$$

and $\delta(s-v)$ is the Dirac δ-function. Hence using (2.1.23)

$$\int (\theta(t) - m(t))\,\bar{\sigma}(t,v)\,dt = \int\int u(s)\,\_\,\sigma(s,t)\,\bar{\sigma}(t,v)\,dt\,ds$$

$$= \int u(s)\,\delta(s-v)\,ds = u(v), \tag{2.1.26}$$

and

$$I(f^*:f) = (1/2)\int\int (\theta(t) - m(t))\,\bar{\sigma}(s,t)(\theta(s) - m(s))\,ds\,dt$$

$$= (1/2) \int\int d(t)\,d(s) \left(\sum_{1}^{\infty} \frac{\alpha_i(s)\,\alpha_i(t)}{\lambda_i} \right) ds\,dt$$

$$= (1/2) \sum_{1}^{\infty} \frac{d_i^2}{\lambda_i} \tag{2.1.27}$$

where

$$d(t) = \theta(t) - m(t), \quad d_i = \int d(t)\,\alpha_i(t)\,dt. \tag{2.1.28}$$

Since, as was remarked, Z in (2.1.15) is a normal random variable, the discrimination information as given by Z is

$$I(1:2\,;Z) = (1/2) \frac{(E_1(Z) - E_2(Z))^2}{\sigma_Z^2}$$

$$= (1/2) \frac{\left(\int d(t)\,u(t)\,dt \right)^2}{\int\int \sigma(s,t)\,u(s)\,u(t)\,ds\,dt}. \tag{2.1.29}$$

But for $u(s)$ satisfying (2.1.23)

$$\int\int \sigma(s,t)\,u(s)\,u(t)\,ds\,dt = \int d(t)\,u(t)\,dt \tag{2.1.30}$$

hence using (2.1.24)

$$I(1:2\,;Z) = (1/2) \int d(t)\,u(t)\,dt = I(f^* : f), \tag{2.1.31}$$

thus Z defined in (2.1.15) is a sufficient linear discriminant functional when $u(s)$ is determined as in (2.1.23).

Note that if we write, using (2.1.1)

$$y_i = \int X(t)\, \alpha_i(t)\, dt = X(\alpha_i) \tag{2.1.32}$$

$$\mathrm{Var}(y_i) = \lambda_i = \int\int \sigma(s,t)\, \alpha_i(s)\, \alpha_i(t)\, ds\, dt \tag{2.1.33}$$

$$\mathrm{Cov}(y_i y_j) = 0 = \int\int \sigma(s,t)\, \alpha_i(s)\, \alpha_j(t)\, ds\, dt \tag{2.1.34}$$

then the y_i correspond to the generalized process as in Gelfand and Yaglom (1957), p. 211, where in this case the space denoted by ϕ in Gelfand and Yaglom (1957) is spanned by the complete set of orthonormal functions $\{a_i(t)\}$. Hence in accordance with Gelfand and Yaglom (1957, (1.29)) (cf (2.1.27)),

$$I(1:2) = (1/2) \sum_{1}^{\infty} \frac{d_i^2}{\lambda_i} \tag{2.1.35}$$

where

$$d_i = E_1(y_i) - E_2(y_i) = \int d(t)\, \alpha_i(t)\, dt \tag{2.1.36}$$

and

$$d(t) = E_1(x(t)) - E_2(x(t)). \tag{2.1.37}$$

The relationship between the y_i and $x(t)$ is that of an orthogonal transformation.

As a particular illustration of the discussion in Example 2.1 let us reexamine the Ornstein-Uhlenbeck process already considered in (2.22) - (2.38) of section 2 of chapter IV , assuming $m_1(t) = m$ under H_1 and $m_2(t) \equiv 0$ under H_2. In this case (2.1.23) is

$$m = \int_0^T e^{-\beta|s-t|} u(s)\, ds \tag{2.1.38}$$

and the solution of (2.1.38) for $u(s)$ is (see Helstrom (1960) , pp. 109 - 114)

$$u(s) = \frac{\delta(s) + \delta(s-T) + \beta}{2}\, m \tag{2.1.39}$$

where δ is the Dirac δ-function .

We verify that (2.1.39) is the solution of (2.1.38) by

$$\frac{m}{2} \int_0^T e^{-\beta|s-t|} (\delta(s) + \delta(s-T) + \beta)\, ds$$

$$= \frac{m}{2} \left\{ e^{-\beta t} + e^{-\beta(T-t)} + \beta \left[\int_0^t e^{-\beta(t-s)}\, ds + \int_t^T e^{-\beta(s-t)}\, ds \right] \right\}$$

$$= \frac{m}{2} \left\{ e^{-\beta t} + e^{-\beta(T-t)} + \beta \left[\frac{e^{-\beta t}(e^{\beta t} - 1)}{\beta} + \frac{e^{\beta t}(e^{-\beta T} - e^{-\beta t})}{-\beta} \right] \right\}$$

$$= \frac{m}{2} \left\{ e^{-\beta t} + e^{-\beta(T-t)} - e^{-\beta t} + 1 - e^{-\beta(T-t)} + 1 \right\}$$

$$= m. \tag{2.1.40}$$

Thus (2.1.24) yields

$$I = (1/2) \int_0^T \frac{\delta(t) + \delta(t-T) + \beta}{2}\, m^2\, dt$$

$$= \frac{m^2}{4} (1 + 1 + \beta T) = \frac{m^2}{2} (1 + \frac{\beta T}{2}) \tag{2.1.41}$$

as given in (2.27) of section 2 of chapter IV. The linear discriminant functional (2.1.15) is now

$$Z = \frac{m}{2}\int_0^T (\delta(t) + \delta(t-T) + \beta)X(t)\,dt$$

$$= \frac{m}{2}(X(0) + X(T) + \beta\int_0^T X(t)\,dt) \tag{2.1.42}$$

which differs by a constant from the log-likelihood ratio given following (2.38) in section 2 of chapter IV. It is seen that

$$E_1(Z) = \frac{m}{2}(m + m + \beta m T) = \frac{m^2}{2}(2 + \beta T),\; E_2(Z) = 0\,, \tag{2.1.43}$$

$$Var(Z) = \int_0^T\int_0^T \frac{m}{2}(\delta(s) + \delta(s-T) + \beta)\frac{m}{2}(\delta(t) + \delta(t-T) + \beta)e^{-\beta|s-t|}\,dsdt$$

$$= \int_0^T \frac{m}{2}(\delta(s) + \delta(s-T) + \beta)\,m\,ds$$

$$= \frac{m^2}{2}(1 + 1 + \beta T) = \frac{m^2}{2}(2 + \beta T), \tag{2.1.44}$$

$$I(1:2;Z) = (1/2)\frac{(E_1(Z) - E_2(Z))^2}{Var(Z)} = (1/2)\frac{\frac{m^4}{4}(2 + \beta T)^2}{\frac{m^2}{2}(2 + \beta T)}$$

$$= \frac{m^2}{2}(1 + \frac{\beta T}{2})\,. \tag{2.1.45}$$

For another approach see Lilliefors (1964).

Example 2.2 Consider the normal stationary random process with discrete time parameter represented by (Borpujari and Kullback (1983))

$$X(t) = \int_{-\pi}^{\pi} \frac{e^{it\lambda}}{(2\pi)^{1/2}}\, dy(\lambda),\ t = \ldots -1, 0, 1, \ldots,$$

$$\text{with} \quad E\,|\,dy(\lambda)|^2 = dF(\lambda)\,. \tag{2.2.1}$$

The moment generating functional of $X(t)$ is

$$M = E\left(exp \sum_t u(t) X(t)\right) = exp\left\{ \sum_t u(t)\, m(t) + \frac{1}{2} \sum_s \sum_t u(s)\, u(t)\, \sigma(s, t)\right\} \tag{2.2.2}$$

or

$$\begin{aligned} ln\ M &= \sum_t u(t)\, m(t) + \frac{1}{2} \sum_s \sum_t u(s)\, u(t)\, \sigma(s, t) \\ &= \int_{-\pi}^{\pi} P_u(e^{i\lambda})\, d\text{v}(\lambda) + \frac{1}{2} \int_{-\pi}^{\pi} |\, P_u(e^{i\lambda})\,|^2\, dF(\lambda) \end{aligned} \tag{2.2.3}$$

where

$$P_u(e^{i\lambda}) = \sum_t \frac{u(t)\, e^{i\lambda t}}{(2\pi)^{1/2}}\,,\ E(X(t)) = m(t) = \int_{-\pi}^{\pi} \frac{e^{it\lambda}}{(2\pi)^{1/2}}\, d\text{v}(\lambda),$$

$$\text{cov}\,(X(s), X(t)) = \sigma(s, t) = \int_{-\pi}^{\pi} \frac{e^{i\lambda(s-t)}}{2\pi}\, dF(\lambda) \tag{2.2.4}$$

and

$$f^*(X(t)) = \frac{exp\left\{ \sum_t u(t) X(t) \right\} f(X(t))}{M} \tag{2.2.5}$$

$$I(f^*:f) = E^*(\sum_t u(t)X(t)) - \ln M$$

$$= E^*\left(\int_{-\pi}^{\pi} P_u(e^{i\lambda})\, dy(\lambda)\right) - \int_{-\pi}^{\pi} P_u(e^{i\lambda})\, d\nu(\lambda)$$

$$- \frac{1}{2}\int_{-\pi}^{\pi} |P_u(e^{i\lambda})|^2\, dF(\lambda)$$

$$= \int_{-\pi}^{\pi} P_u(e^{i\lambda})\, d\theta(\lambda) - \int_{-\pi}^{\pi} P_u(e^{i\lambda})\, d\nu(\lambda) - \frac{1}{2}\int_{-\pi}^{\pi} |P_u(e^{i\lambda})|^2\, dF(\lambda) \tag{2.2.6}$$

$$\theta^*(t) = \int_{-\pi}^{\pi} \frac{e^{it\lambda}}{(2\pi)^{1/2}}\, d\theta(\lambda) = E^*(X(t))$$

$$= E\left(\frac{X(t)\exp\left\{\sum_t u(t)X(t)\right\}}{M}\right)$$

$$= \frac{E(X(t)\exp\left\{\int_{-\pi}^{\pi} P_u(e^{i\lambda})\, dy(\lambda)\right\})}{M} \tag{2.2.7}$$

$$\sum_t u(t)\theta^*(t) = \int_{-\pi}^{\pi} P_u(e^{i\lambda})\, d\theta(\lambda)$$

$$= \frac{E\left(\left(\int_{-\pi}^{\pi} P_u(e^{i\lambda}) dy(\lambda)\right)\exp\left\{\int_{-\pi}^{\pi} P_u(e^{i\lambda}) dy(\lambda)\right\}\right)}{M} = E(Ze^Z)/E(e^Z) \tag{2.2.8}$$

where

$$Z = \int_{-\pi}^{\pi} P_u (e^{i\lambda})\, dy(\lambda) \tag{2.2.9}$$

or

$$Z = \sum_t u(t) X(t), \; \sigma_Z^2 = \sum_t \sum_s u(t) u(s)\, \sigma(s-t)$$

$$= \int_{-\pi}^{\pi} |P_u (e^{i\lambda})|^2 \, dF(\lambda). \tag{2.2.10}$$

We note that Z is a normally distributed random variable since it is a linear functional of the normal process $X(t)$. It is found that

$$E(exp\, \tau Z) = exp \left\{ \tau \int_{-\pi}^{\pi} P_u (e^{i\lambda})\, d\nu(\lambda) + \frac{\tau^2}{2} \int_{-\pi}^{\pi} |P_u (e^{i\lambda})|^2 \, dF(\lambda) \right\} \tag{2.2.11}$$

$$E(Z\, exp\, \tau Z) = E(exp\, \tau Z) \left\{ \int_{-\pi}^{\pi} P_u (e^{i\lambda})\, d\nu(\lambda) \right.$$

$$\left. + \tau \int |P_u (e^{i\lambda})|^2 \, dF(\lambda) \right\} \tag{2.2.12}$$

$$E(Z e^Z) / E(e^Z) = \int_{-\pi}^{\pi} P_u (e^{i\lambda})\, d\nu(\lambda) + \int |P_u (e^{i\lambda})|^2 \, dF(\lambda) \tag{2.2.13}$$

and using (2.2.8), (2.2.6), and (2.2.14)

$$\sum_t u(t) \theta^*(t) = \int_{-\pi}^{\pi} P_u (e^{i\lambda})\, d\theta(\lambda) = \int_{-\pi}^{\pi} P_u (e^{i\lambda})\, d\nu(\lambda)$$

$$+ \int_{-\pi}^{\pi} | P_u(e^{i\lambda}) |^2 \, dF(\lambda) \tag{2.2.14}$$

$$I(f^* : f) = \frac{1}{2} \int_{-\pi}^{\pi} | P_u(e^{i\lambda}) |^2 \, dF(\lambda) \tag{2.2.15}$$

$$\int_{-\pi}^{\pi} P_u(e^{i\lambda}) \left\{ d(\theta(\lambda) - v(\lambda)) - P_u(e^{-i\lambda}) \, dF(\lambda) \right\} = 0 . \tag{2.2.16}$$

Hence, since $\{e^{it\lambda}\}$, $t = \ldots -1, 0, 1, \ldots$ is a complete orthonormal set

$$d(\theta(\lambda) - v(\lambda)) = P_u(e^{-i\lambda}) \, dF(\lambda) \tag{2.2.17}$$

$$I(f^* : f) = \frac{1}{2} \int_{-\pi}^{\pi} \frac{| d(\theta(\lambda) - v(\lambda)) |^2}{dF(\lambda)} . \tag{2.2.18}$$

Note the similarity of the expression in (2.1.27) and in (2.2.18) .

Let us now apply the technique used in section 2 of chapter IV to find the log- likelihood. We recall that here

$$H_1 : E_1(X(t)) = \theta^*(t) = \int_{-\pi}^{\pi} \frac{e^{it\lambda}}{(2\pi)^{1/2}} \, d\theta(\lambda) ,$$

$$H_2 : E_2(X(t)) = m(t) = \int_{-\pi}^{\pi} \frac{e^{it\lambda}}{(2\pi)^{1/2}} \, dv(\lambda) . \tag{2.2.19}$$

Using (2.2.18) it is found that

$$\hat{I}\,(*:H_2) - \hat{I}\,(*:H_1) = \frac{1}{2}\int_{-\pi}^{\pi} \frac{|d(y(\lambda) - v(\lambda))|^2}{dF(\lambda)}$$

$$-\frac{1}{2}\int_{-\pi}^{\pi} \frac{|d(y(\lambda) - \theta(\lambda))|^2}{dF(\lambda)}$$

$$= \frac{1}{2}\int_{-\pi}^{\pi} \frac{dy(\lambda)\,.\,d(\overline{\theta(\lambda)} - \overline{v(\lambda)}) + \overline{dy(\lambda)}\,.\,d(\theta(\lambda) - v(\lambda))}{dF(\lambda)}$$

$$+ \frac{1}{2}\int_{-\pi}^{\pi} \frac{|dv(\lambda)|^2 - |d\theta(\lambda)|^2}{dF(\lambda)}, \tag{2.2.20}$$

where the bar over an expression denotes the complex conjugate. Using (2.2.17) in (2.2.20) the log- likelihood may also be written as

$$\ell(X(t)) = \frac{1}{2}\int_{-\pi}^{\pi} (P_u(e^{i\lambda})\,\overline{dy(\lambda)} + P_u(e^{-i\lambda})\,dy(\lambda))$$

$$+ \frac{1}{2}\int_{-\pi}^{\pi} \frac{|dv(\lambda)|^2 - |d\theta(\lambda)|^2}{dF(\lambda)}. \tag{2.2.21}$$

If $X(t)$ is a real process, that is, (2.2.1) may also be written as

$$X(t) = \int_{-\pi}^{\pi} \frac{e^{-it\lambda}\,\overline{dy(\lambda)}}{(2\pi)^{1/2}}, \tag{2.2.22}$$

then (2.2.21) may be written as

$$\ell(X(t)) = \int_{-\pi}^{\pi} P_u(e^{i\lambda})\,dy(\lambda) + \frac{1}{2}\int_{-\pi}^{\pi} \frac{|dv(\lambda)|^2 - |d\theta(\lambda)|^2}{dF(\lambda)} \tag{2.2.23}$$

and we see that the linear discriminant functional Z in (2.2.9) or (2.2.10) subject to (2.2.17) differs from the log-likelihood by a constant term.

We make the following remarks. It was seen that Z defined in (2.1.15) is a sufficient linear discriminant functional when $u(s)$ is determined as in (2.1.23). We then note from (2.1.29) and (2.1.31) that if $E_1(Z) - E_2(Z)$ is finite, but

$I(f^* : f) = \infty$, then $\sigma_Z^2 = 0$ and there is perfect discrimination.

Similar results hold for Z in (2.2.10) with $u(t)$ satisfying (2.2.17). See Anorina (1970).

For a proof of the maximum entropy characterization of autoregressive spectral densities as models for the spectral density of a stationary time series see Parzen (1982).

Example 2.3 This example is closely related to Example 2.2 and we shall omit some of the details leaving them to the reader. Consider the normal stationary random process represented by

$$X(t) = \int_{-\infty}^{\infty} \frac{e^{it\lambda}}{(2\pi)^{1/2}}\, dy(\lambda), \quad \mathrm{cov}(X(s), X(t)) = \int_{-\infty}^{\infty} \frac{e^{i\lambda(s-t)}}{2\pi}\, dF(\lambda) \tag{2.3.1}$$

with moment generating functional

$$M = E\left\{ exp \int u(t) X(t)\, dt \right\} = E\left\{ exp \int_{-\infty}^{\infty} \int \frac{e^{it\lambda}}{(2\pi)^{1/2}} u(t)\, dy(\lambda)\, dt \right\}$$

$$= E\left\{ exp \int U(\lambda)\, dy(\lambda) \right\} \tag{2.3.2}$$

where

$$U(\lambda) = \int \frac{e^{it\lambda}}{(2\pi)^{1/2}} u(t)\, dt. \tag{2.3.3}$$

It is found that

$$ln\, M = \int U(\lambda)\, d\mathrm{v}(\lambda) + \frac{1}{2} \int |U(\lambda)|^2\, dF(\lambda) \tag{2.3.4}$$

$$f^*(X(t)) = \frac{exp\left\{\int u(t)\, X(t)\, dt\right\} f(X(t))}{M} \tag{2.3.5}$$

$$I(f^* : f) = \int U(\lambda)\, d\theta(\lambda) - \int U(\lambda)\, d\mathrm{v}(\lambda) - \frac{1}{2} \int |U(\lambda)|^2\, dF(\lambda), \tag{2.3.6}$$

and in a manner similar to that in Example 2.2

$$I(f^* : f) = \frac{1}{2} \int |U(\lambda)|^2\, dF(\lambda) \tag{2.3.7}$$

$$\int U(\lambda)[\, d(\theta(\lambda) - \mathrm{v}(\lambda)) - \overline{U}(\lambda)\, dF(\lambda)] = 0 \tag{2.3.8}$$

$$d(\theta(\lambda) - \mathrm{v}(\lambda)) = \overline{U}(\lambda)\, dF(\lambda) \tag{2.3.9}$$

$$I(f^* : f) = \frac{1}{2} \int \frac{|d(\theta(\lambda) - \mathrm{v}(\lambda))|^2}{dF(\lambda)} \tag{2.3.10}$$

where $\overline{U}(\lambda)$ is the complex conjugate of $U(\lambda)$.

In a manner similar to that in example 2.2 it is found that

$$\hat{I}\,(*:H_2) - \hat{I}\,(*:H_1) = \tag{2.3.11}$$

$$= \tfrac{1}{2}\int \frac{dy(\lambda)\cdot \overline{d\,(\theta(\lambda)} - \overline{v(\lambda)}\,) + \overline{dy(\lambda)}\cdot d(\theta(\lambda) - v(\lambda))}{dF(\lambda)} +$$

$$+ \tfrac{1}{2}\int \frac{|d\,v(\lambda)|^2 - |d\,\theta(\lambda)|^2}{dF(\lambda)}$$

and using (2.3.9) the log-likelihood may be written as

$$\ell(X(t)) = \tfrac{1}{2}\int (U(\lambda)\; dy(\lambda) + \overline{U(\lambda)}\;\; \overline{dy(\lambda)}\,) + \tfrac{1}{2}\int \frac{|dv(\lambda)|^2 - |d\theta(\lambda)|^2}{dF(\lambda)}\,. \tag{2.3.12}$$

If $X(t)$ is a real process, that is (2.3.1) may also be written as

$$X(t) = \int_{-\infty}^{\infty} \frac{e^{-it\lambda}}{(2\pi)^{\frac{1}{2}}}\; \overline{dy(\lambda)}\,, \tag{2.3.13}$$

then (2.3.12) may be written as

$$\int U(\lambda)\, dy(\lambda) + \tfrac{1}{2}\int \frac{|dv(\lambda)|^2 - |d\theta(\lambda)|^2}{dF(\lambda)} \tag{2.3.14}$$

and we note that the linear discriminant functional (see Shumway (1965, p. 43))

$$Z = \int u(t)\, X(t)\, dt = \int U(\lambda)\, dy(\lambda) \tag{2.3.15}$$

subject to (2.3.9) differs from the log-likelihood by a constant term.

If we assume that under $H_2 : E_2(X(t)) \equiv 0$, that is, $v(\lambda) \equiv 0$, and under H_1 : $E_1(X(t)) = m_1(t)$, where

$$m_1(t) = \int \frac{e^{it\lambda}}{(2\pi)^{\frac{1}{2}}} \, d\theta(\lambda), \tag{2.3.16}$$

then (2.3.9) implies

$$m_1(t) = \int \frac{e^{it\lambda}}{(2\pi)^{\frac{1}{2}}} \, \overline{U(\lambda)} \, dF(\lambda) \tag{2.3.17}$$

and the log-likelihood in (2.3.14) may be written as

$$\ell(X(t)) = \int U(\lambda)\, dy(\lambda) - \tfrac{1}{2} \int |U(\lambda)|^2 dF(\lambda) . \tag{2.3.18}$$

See Yaglom in Rosenblatt (1963, p. 332).

Example 2.4 (See Example 2.1). Consider the case of white noise limited to the band $(0, W)$ cycles per second (see Rosenblatt (1962, p. 179), Rozanov (1963, (6.22))

$$X(t) = \sum_{n=-\infty}^{\infty} X_n \frac{\sin \pi (2Wt - n)}{\pi (2Wt - n)}, \quad X_n = X\left(\frac{n}{2W}\right) \tag{2.4.1}$$

where X_n is $N(0, \sigma^2)$, and the covariance function is

$$\begin{aligned} E(X(s)X(t)) = \sigma(s,t) &= \sum_{n=-\infty}^{\infty} \sigma^2 \frac{\sin \pi (2Ws - n)}{\pi (2Ws - n)} \frac{\sin \pi (2Wt - n)}{\pi (2Wt - n)} \\ &= \sigma^2 \frac{\sin 2\pi W(s-t)}{2\pi W(s-t)} . \end{aligned} \tag{2.4.2}$$

Let us set $y_n = X_n / (2W)^{\frac{1}{2}}$ and $a_n(t) = (2W)^{\frac{1}{2}} \sin \pi (2Wt - n) / \pi (2Wt - n)$ so that (2.4.1) may be written in a form similar to (2.1.1) as

$$X(t) = \sum_{n=-\infty}^{\infty} y_n \alpha_n(t) \tag{2.4.3}$$

where, as we shall show, the $\{\alpha_n(t)\}$ are an orthonormal set of functions and $\mathrm{var}(y_n) = \sigma^2 / 2W$.

Suppose now that the process (signal) $Y(t)$ is given as in (2.4.4) under the hypotheses H_2 and H_1 respectively,

$$H_2 : Y(t) = X(t) \tag{2.4.4}$$

$$H_1 : Y(t) = X(t) + S(t)$$

where S(t) is a non-random process (signal) such that

$$S(t) = \sum_{n=-\infty}^{\infty} s_n \frac{\sin \pi (2Wt - n)}{\pi (2Wt - n)}, \; s_n = S\left(\frac{n}{2W}\right) \tag{2.4.5}$$

$$= \sum_{n=-\infty}^{\infty} \frac{s_n}{(2W)^{\frac{1}{2}}} \alpha_n(t)$$

and hence, the covariance function of Y(t) is

$$\sigma_y(s, t) = \sum_{n=-\infty}^{\infty} \frac{\sigma^2}{2W} \alpha_n(s) \alpha_n(t) \tag{2.4.6}$$

under both hypotheses.

In this case it is found that (2.1.27) becomes

$$I(f^* : f) = \tfrac{1}{2} \int \int S(t) S(s) \bar{\sigma}(s, t) \, ds \, dt =$$

$$= \tfrac{1}{2} \int \int S(t) S(s) \left(\sum_n \left(\frac{\alpha_n(s) \alpha_n(t) 2W}{\sigma^2} \right) dsdt \right. \tag{2.4.7}$$

$$= \tfrac{1}{2} \iint \Big(\sum_j \frac{s_j}{(2W)^{\frac{1}{2}}} \alpha_j(t)\Big) \left(\sum_k \frac{s_k}{(2W)^{\frac{1}{2}}} \alpha_k(s)\right)$$

$$\left(\sum_n \frac{\alpha_n(s)\alpha_n(t)}{\sigma^2} 2W\right) ds dt$$

$$= \frac{1}{2\sigma^2} \int \Big(\sum_j s_j \alpha_j(t)\Big)\Big(\sum_n s_n \alpha_n(t)\Big) dt$$

$$= \frac{1}{2\sigma^2} \sum_n s_n^2 = \frac{2W}{2\sigma^2} \int (S(t))^2 dt$$

and (2.1.28) yields

$$d(t) = S(t), \; d_i = s_i/(2W)^{\frac{1}{2}} \tag{2.4.8}$$

so that

$$I(f^* : f) = \tfrac{1}{2} \sum \frac{d_i^2}{\lambda_i} = \frac{2W}{2\sigma^2} \sum s_i^2 / 2W = \frac{1}{2\sigma^2} \sum_n s_n^2 \tag{2.4.9}$$

as in (2.4.7).

Note that here the linear discriminant functional is given by

$$Z = \int u(t) Y(t) dt \tag{2.4.10}$$

and from (2.1.23)

$$S(t) = \int u(s) \Big(\sum_n \frac{\sigma^2}{2W} \alpha_n(s)\alpha_n(t)\Big) ds = \frac{\sigma^2}{2W} \sum_n \alpha_n(t) \int u(s) \alpha_n(s) ds$$

$$= \frac{\sigma^2}{2W} \sum_n \alpha_n(t) u_n = \frac{\sigma^2}{2W} u(t), \tag{2.4.11}$$

that is

$$Z = \frac{2W}{\sigma^2} \int S(t)\, Y(t)\, dt. \tag{2.4.12}$$

Note that $E_2(Z) = 0,\ E^*(Z) = \dfrac{2W}{\sigma^2} \displaystyle\int (S(t))^2\, dt,$

$$\sigma_Z^2 = 4W^2 \iint \frac{S(t)\, S(s)}{\sigma^4}\, \sigma_y(s,t)\, ds\, dt = \frac{2W}{\sigma^2} \int (S(t))^2\, dt.$$

Let us apply here the technique used in section 2 of Chapter IV to find the log-likelihood. Using (2.4.7) it is found that

$$\begin{aligned} \hat{I}(*:H_2) - \hat{I}(*:H_1) &= \frac{2W}{2\sigma^2} \int (Y(t))^2\, dt - \frac{2W}{2\sigma^2} \int (Y(t) - S(t))^2\, dt \\ &= \frac{2W}{\sigma^2} \int (Y(t)\, S(t)\, dt - \frac{2W}{2\sigma^2} \int (S(t))^2\, dt \end{aligned} \tag{2.4.13}$$

leading to the same discriminant functional as in (2.4.12).

As in (2.1.32) - (2.1.37) we may define a generalized process and related values by

$$y_n = \int X(t)\, (2W)^{\frac{1}{2}}\, (\sin \pi (2Wt - n) / \pi (2Wt - n))\, dt \tag{2.4.14}$$

$$\begin{aligned} \mathrm{Var}(y_n) = \frac{\sigma^2}{2W} &= \iint \sigma^2 \frac{\sin 2\pi W(s-t)}{2\pi W(s-t)} (2W)^{\frac{1}{2}} \frac{\sin \pi (2Ws - n)}{\pi (2Ws - n)} \\ &\quad (2W)^{\frac{1}{2}} \frac{\sin \pi (2Wt - n)}{\pi (2Wt - n)}\, ds dt \end{aligned} \tag{2.4.15}$$

$$\mathrm{Cov}(y_m, y_n) = 0 = \iint \sigma^2 \frac{\sin 2\pi W(s-t)}{2\pi W(s-t)} (2W)^{\frac{1}{2}} \frac{\sin \pi (2Ws - m)}{\pi (2Ws - m)} \tag{2.4.16}$$

$$(2W)^{\frac{1}{2}} \frac{\sin \pi (2Wt - n)}{\pi (2Wt - n)} \, dsdt$$

$$d_n = E_1(y_n) - E_2(y_n) = \int S(t) (2W)^{\frac{1}{2}} \frac{\sin \pi (2Wt - n)}{\pi (2Wt - n)} \, dt \tag{2.4.17}$$

$$= s_n / (2W)^{\frac{1}{2}}$$

$$I(1:2) = \tfrac{1}{2} \sum \frac{s_n^2 / 2W}{\sigma^2 / 2W} = \frac{1}{2\sigma^2} \sum s_n^2 \tag{2.4.18}$$

$$= \frac{2W}{2\sigma^2} \int (S(t))^2 \, dt$$

as in (2.4.7) .

We shall now show that the $\{\alpha_n(t)\}$ as in (2.4.3) form an orthonormal set. Since

$$\alpha_n(t) = (2W)^{\frac{1}{2}} \frac{\sin \pi (2Wt - n)}{\pi (2Wt - n)} = \tag{2.4.19}$$

$$= \int_{-2\pi W}^{2\pi W} \frac{(2W)^{\frac{1}{2}}}{4\pi W} e^{\frac{-i\lambda n}{2W}} e^{i\lambda t} \, d\lambda$$

$$\int_{-\infty}^{\infty} \alpha_m(t) \alpha_n(t) \, dt = \int_{-\infty}^{\infty} dt \int_{-2\pi W}^{2\pi W} \frac{(2W)^{\frac{1}{2}}}{4\pi W} e^{-i\lambda \frac{m}{2W} + i\lambda t} \, d\lambda \tag{2.4.20}$$

$$\int_{-2\pi W}^{2\pi W} \frac{(2W)^{\frac{1}{2}}}{4\pi W} e^{-i\mu \frac{n}{2W} + i\mu t} \, d\mu$$

$$= \int_{-2\pi W}^{2\pi W} \int_{-2\pi W}^{2\pi W} \frac{2W}{(4\pi W)^2} e^{-i\lambda \frac{m}{2W} - i\mu \frac{n}{2W}} 2\pi\delta(\lambda + \mu)\, d\lambda\, d\mu$$

$$= \frac{1}{4\pi W} \int_{-2\pi W}^{2\pi W} e^{\frac{-i\lambda}{2W}(m-n)}\, d\lambda = \frac{\sin\pi(n-m)}{\pi(n-m)} = \begin{matrix} 0 & n \neq m \\ 1 & n = m \end{matrix} .$$

If we set

$$A_n(\lambda) = \frac{(2W)^{\frac{1}{2}}}{4\pi W} \exp(-i\lambda n / 2W) \tag{2.4.21}$$

then we have

$$a_n(t) = \int_{-2\pi W}^{2\pi W} A_n(\lambda)\, e^{i\lambda t}\, d\lambda \tag{2.4.22}$$

$$A_n(\lambda) = \frac{1}{2\pi} \int_{-\infty}^{\infty} e^{-i\lambda t}\, a_n(t)\, dt \tag{2.4.23}$$

and applying Parseval's theorem we get

$$\int_{-\infty}^{\infty} a_m(t)\, a_n(t)\, dt = 2\pi \int_{-2\pi W}^{2\pi W} A_m(\lambda)\, \overline{A_n(\lambda)}\, d\lambda \tag{2.4.24}$$

which is the result in (2.4.20).

Let us reconsider this example using the approach as in example 2.3 where we consider all λ functions as zero outside the interval $-2\pi W \leq \lambda \leq 2\pi W$. Thus

$$X(t) = \int_{-2\pi W}^{2\pi W} e^{i\lambda t}\, dy(\lambda) \tag{2.4.25}$$

$$\mathrm{Cov}(X(s), X(t)) = \sigma^2 \frac{\sin 2\pi W(s-t)}{2\pi W(s-t)} = \int_{-2\pi W}^{2\pi W} e^{i\lambda(s-t)}\, dF(\lambda) \tag{2.4.26}$$

so that $dF(\lambda) = \sigma^2 \, d\lambda / 4\pi W$ for $-2\pi W \le \lambda \le 2\pi W$ and zero elsewhere.

We now consider the hypotheses in (2.4.4) with

$$S(t) = \int_{-2\pi W}^{2\pi W} e^{i\lambda t} \, d\theta(\lambda) \; . \tag{2.4.27}$$

The result in (2.3.10) becomes

$$I(f^* : f) = \frac{4\pi W}{2\sigma^2} \int_{-2\pi W}^{2\pi W} \frac{|d\theta(\lambda)|^2}{d\lambda} = \frac{4\pi W}{2\sigma^2} \int_{-2\pi W}^{2\pi W} \left|\frac{d\theta(\lambda)}{d\lambda}\right|^2 d\lambda \; . \tag{2.4.28}$$

Using Parseval's theorem, (2.4.27) yields

$$\int (S(t))^2 \, dt = 2\pi \int_{-2\pi W}^{2\pi W} \left|\frac{d\theta(\lambda)}{d\lambda}\right|^2 d\lambda \tag{2.4.29}$$

and hence (2.4.28) becomes

$$I(f^* : f) = \frac{2W}{2\sigma^2} \int (S(t))^2 \, dt \tag{2.4.30}$$

as in (2.4.7).

The condition for the sufficient linear discriminant functional given by (2.3.9) yields

$$\overline{U}(\lambda) = \frac{d\theta(\lambda)}{dF(\lambda)} = \frac{4\pi W}{\sigma^2} \frac{d\theta(\lambda)}{d\lambda} \, , \; -2\pi W \le \lambda \le 2\pi W \tag{2.4.31}$$

and the linear discriminant functional (2.3.15) is

$$Z = \frac{4\pi W}{\sigma^2} \int_{-2\pi W}^{2\pi W} \frac{d\theta(\lambda)}{d\lambda} \, dy(\lambda) = \frac{2W}{\sigma^2} \int S(t) \, Y(t) \, dt \tag{2.4.32}$$

using Parseval's theorem and getting the result in (2.4.12). The log-likelihood in (2.3.18) is now

$$\ell(X(t)) = \frac{4\pi W}{\sigma^2} \int_{-2\pi W}^{2\pi W} \frac{d\theta}{d\lambda}\, dy(\lambda) - \tfrac{1}{2}\left(\frac{4\pi W}{\sigma^2}\right)^2 \int_{-2\pi W}^{2\pi W} \left|\frac{d\theta(\lambda)}{d\lambda}\right|^2 \frac{\sigma^2\, d\lambda}{4\pi W}$$

$$= \frac{2W}{\sigma^2} \int Y(t)\, S(t)\, dt - \frac{2W}{2\sigma^2} \int (S(t))^2\, dt \tag{2.4.33}$$

again using Parseval's theorem and getting the result in (2.4.13).

It is to be noted that the total energy in the signal is

$$E = \int S(t)^2\, dt = \sum_{n=-\infty}^{\infty} \frac{s_n^2}{2W} \tag{2.4.34}$$

and the mean noise power per unit bandwidth is

$$N = 2\pi\,(f(\lambda) + f(-\lambda)) = 2\pi\left(\frac{\sigma^2}{4\pi W} + \frac{\sigma^2}{4\pi W}\right) = \frac{\sigma^2}{W} \tag{2.4.35}$$

where, as in (2.4.26), $dF(\lambda) = \sigma^2\, d\lambda/4\pi W = f(\lambda)\, d\lambda$. Thus (2.4.18) may be written as

$$I(1:2) = E/N \tag{2.4.36}$$

where E/N is the signal to noise ratio (cf. Kullback (1959, p. 9)).

Example 2.5 For linear Brownian motion or the Wiener process (see (2.39) in Chapter IV, Rosenblatt (1962, p. 195))

$$X(t) = \left(Z_0 t + \sum_{n=1}^{\infty} Z_n \frac{\sin n t\pi}{n\pi}\right)(2)^{\frac{1}{2}}\ , \quad 0 < t < 1 \tag{2.5.1}$$

where the Z_n's are $N(0, \sigma^2)$, $n = 1, 2, \ldots$ and Z_o is $N(0, \sigma^2/2)$. The covariance function is

$$E(X(s)X(t)) = \sigma^2 \min(s, t)$$
$$= \sigma^2 \left(st + \frac{2}{\pi^2} \sum_{n=1}^{\infty} \frac{\sin n\pi s \sin n\pi t}{n^2} \right). \tag{2.5.2}$$

We may verify (2.5.2) by writing

$$st + \frac{2}{\pi^2} \sum_{n=1}^{\infty} \frac{\sin n\pi s \sin n\pi t}{n^2} \tag{2.5.3}$$
$$= st + \frac{1}{\pi^2} \{ \sum_{n=1}^{\infty} \frac{\cos n(s-t)\pi}{n^2} - \sum_{n=1}^{\infty} \frac{\cos n(s+t)\pi}{n^2} \}.$$

But (Carslaw (1930), p. 265, problem 4)

$$\sum_{n=1}^{\infty} \frac{\cos n\theta}{n^2} = \tfrac{1}{4}(\theta - \pi)^2 - (1/3)\pi^2, \; 0 \le \theta \le 2\pi \tag{2.5.4}$$

hence for $s > t$ (2.5.3) yields

$$st + \frac{1}{\pi^2} \{ \frac{[(s-t)\pi - \pi]^2}{4} - \frac{\pi^2}{3} - \frac{[(s+t)\pi - \pi]^2}{4} + \frac{\pi^2}{3} \} \tag{2.5.5}$$
$$= st + \frac{(s-1-t)^2 - (s-1+t)^2}{4} = st - st + t = t = \min(s, t).$$

Suppose the process $Y(t)$ is defined respectively under the hypotheses H_1 and H_2 as

$$H_1 : Y(t) = X(t) + m(t) \tag{2.5.6}$$
$$H_2 : Y(t) = X(t)$$

where

$$m(t) = \left(m_0 t + \sum_{n=1}^{\infty} m_n \frac{\sin nt\pi}{n\pi} \right) (2)^{\frac{1}{2}} . \tag{2.5.7}$$

Here

$$I(f^* : f) = \frac{1}{2} \sum_{n=1}^{\infty} \frac{m_n^2}{\sigma^2} + \frac{m_o^2}{\sigma^2} . \tag{2.5.8}$$

But $m(t) = \int_0^t m'(t)\,dt$, where

$$m'(t) = \left(m_0 + \tfrac{1}{2} \sum_{n=1}^{\infty} m_n \exp \pi int + \tfrac{1}{2} \sum_{n=-1}^{-\infty} m_n \exp \pi int \right) (2)^{\frac{1}{2}} \tag{2.5.9}$$

and hence by the Parseval theorem

$$\frac{m_o^2}{\sigma^2} + \frac{1}{2} \sum_{n=1}^{\infty} \frac{m_n^2}{\sigma^2} = \frac{1}{2\sigma^2} \int_0^1 (m'(t))^2\,dt \tag{2.5.10}$$

a result already found for the Wiener process as a Markov process (see (2.45) in section 2 of Chapter IV).

This seems to imply that since the "derivative" of $X(t)$ is white noise, the linear operation of differentiation leaves the information unchanged, since the final result is the white noise discrimination information value. To further amplify the last statement let $Z(t)$ be white noise, that is, a normal or Gaussian process with $E(Z(t)) = 0$ and $\text{cov}(Z(s), Z(t)) = \sigma^2 \delta(s - t)$ or spectral distribution $dF_Z(\lambda) = \sigma^2 d\lambda/2\pi$. Let us now consider the process defined by

$$X(t) = \int g(t, u) Z(u)\, du$$

$$\operatorname{cov}(X(s), X(t)) = \int \int g(s, u) g(t, v) \sigma^2 \delta(u - v)\, du\, dv. \tag{2.5.11}$$

Defining $g(t, u)$ as

$$\begin{aligned} g(t, u) &= 1,\ a \le u \le t \le b \\ &= 0, \text{elsewhere} \end{aligned} \tag{2.5.12}$$

it is found that

$$X(t) = \int_a^t Z(u)\, du, \quad \operatorname{cov}(X(s), X(t)) = \sigma^2 \min(s, t). \tag{2.5.13}$$

Hence under H_1 in (2.5.6) we have

$$Y(t) = \int_a^t Z(u)\, du + \int_a^t S(u)\, du \tag{2.5.14}$$

where $m(t) = \int_a^t S(u)\, du$, and $S(u)$ is the signal function as in example 2.4 or example 2.8. Thus the discrimination information for the linear Brownian motion or Wiener process for $a = 0, b = 1$ is

$$\frac{1}{2\sigma^2} \int_0^1 (m'(t))^2\, dt = \frac{1}{2\sigma^2} \int_0^1 (S(t))^2\, dt \tag{2.5.15}$$

where the right-hand side is the same as (2.4.7) or (2.8.6).

As in example 2.4 we may also use here the technique applied in section 2 of Chapter IV to find the log-likelihood ratio. Using (2.5.8) it is found that

$$\hat{I}(*:H_2) - \hat{I}(*:H_1) = \frac{Z_o^2}{\sigma^2} + \tfrac{1}{2}\sum_{n=1}^{\infty}\frac{Z_n^2}{\sigma^2} - \frac{(Z_o - m_o)^2}{\sigma^2} - \sum_{n=1}^{\infty}\frac{(Z_n - m_n)^2}{\sigma^2} \tag{2.5.16}$$

$$= \frac{2Z_o m_o}{\sigma^2} + \sum_{n=1}^{\infty}\frac{Z_n m_n}{\sigma^2} - \frac{m_o^2}{\sigma^2} - \tfrac{1}{2}\sum_{n=1}^{\infty}\frac{m_n^2}{\sigma^2}$$

and the Parseval theorem yields

$$\hat{I}(*:H_2) - \hat{I}(*:H_1) = \frac{1}{\sigma^2}\int_0^1 m'(t)\,dX(t) - \frac{1}{2\sigma^2}\int_0^1 m'(t)^2\,dt, \tag{2.5.17}$$

agreeing with log-likelihood ratio in (2.50) of Chapter IV (when $a = 0$ the first term in (2.50) is zero).

Example 2.6 See Example 2.3 . Suppose we start with the relation (2.1.27), that is

$$I(f^* : f) = \frac{1}{2}\int\int d(t)\,\overline{d(s)}\,\bar{\sigma}(s,t)\,ds\,dt \tag{2.6.1}$$

where $\overline{d(s)}$ is the complex conjugate of $d(s)$ and $\bar{\sigma}(s, t)$ is the inverse function of $\sigma(s, t)$. For $\sigma(s, t)$ in (2.3.1), that is

$$\sigma(s,t) = \sigma(s-t) = \int_{-\infty}^{\infty}\frac{e^{i\lambda(s-t)}}{2\pi}\,dF(\lambda) \tag{2.6.2}$$

it is found that (see Shumway (1965), p. 39, Chapter VI (1.5))

$$\bar{\sigma}(t-u) = \int_{-\infty}^{\infty}\frac{e^{i\lambda(t-u)}(d\lambda)^2}{2\pi\,dF(\lambda)} \tag{2.6.3}$$

since

$$\int \sigma(s-t)\,\bar{\sigma}(t-u)\,dt = \int\int\int e^{i\lambda(s-t)} \frac{dF(\lambda)\cdot e^{i\lambda'(t-u)}(d\lambda')^2}{(2\pi)^2\,dF(\lambda')}\,dt$$

$$= \int\int \frac{e^{i\lambda s - i\lambda' u}\,dF(\lambda)(d\lambda')^2}{2\pi\,dF(\lambda')} \cdot \frac{1}{2\pi}\int e^{-i(\lambda-\lambda')t}\,dt \tag{2.6.4}$$

$$= \int\int \frac{e^{i\lambda s - i\lambda' u}\,dF(\lambda)(d\lambda')^2}{2\pi\,dF(\lambda')}\,\delta(\lambda-\lambda')$$

$$= \int e^{i\lambda(s-u)} \frac{dF(\lambda)}{2\pi\,dF(\lambda)}\,d\lambda = \delta(s-u)\,.$$

Using (2.6.3) in (2.6.1) there is obtained

$$I(f^*:f) = \frac{1}{2}\int\int\int\int \frac{e^{i\lambda s}}{(2\pi)^{\frac{1}{2}}}\,d(\theta(\lambda)-\nu(\lambda))\,\frac{e^{-i\lambda' t}}{(2\pi)^{\frac{1}{2}}}\,d\,\overline{(\theta(\lambda')-\nu(\lambda'))} \cdot \frac{e^{i\lambda''(t-s)}(d\lambda'')^2}{2\pi\,dF(\lambda'')}\,ds\,dt \tag{2.6.5}$$

$$= \frac{1}{2}\int\int\int\int e^{i\lambda s}\,d(\theta(\lambda)-\nu(\lambda))\,d\,\overline{(\theta(\lambda')-\nu(\lambda'))}\,\frac{e^{-i\lambda'' s}(d\lambda'')^2}{2\pi\,dF(\lambda'')}\,\delta(\lambda'-\lambda'')\,ds$$

$$= \frac{1}{2}\int\int\int e^{i\lambda s}\,d(\theta(\lambda)-\nu(\lambda))\,d\,\overline{(\theta(\lambda')-\nu(\lambda'))}\,\frac{e^{-i\lambda' s}\,d\lambda'}{2\pi\,dF(\lambda')}\,ds$$

$$= \frac{1}{2}\int\int d(\theta(\lambda)-\nu(\lambda))\,d\,\overline{(\theta(\lambda')-\nu(\lambda'))}\,\frac{d\lambda'}{dF(\lambda')}\,\delta(\lambda-\lambda')$$

$$= \frac{1}{2}\int \frac{|d(\theta(\lambda)-\nu(\lambda))|^2}{dF(\lambda)}$$

as in (2.3.10).

Example 2.7 Consider the Ornstein-Uhlenbeck process as in Chapter IV (2.22). This is a normal, stationary random process as in example 2.3 with $\sigma(s-t) = exp(-\beta|t-s|)$. From (2.3.1), that is

$$e^{-\beta|t-s|} = \int_{-\infty}^{\infty} \frac{e^{i\lambda(s-t)}}{2\pi} \, dF(\lambda) \tag{2.7.1}$$

it is found that

$$dF(\lambda) = 2\beta d\lambda / (\lambda^2 + \beta^2). \tag{2.7.2}$$

As in Chapter IV (2.28) let us assume that

$$E_1(X(t)) = m(t), \; E_2(X(t)) = 0 \tag{2.7.3}$$

and also that

$$m(t) = \int \frac{e^{it\lambda}}{(2\pi)^{\frac{1}{2}}} M(\lambda)\, d\lambda . \tag{2.7.4}$$

Using (2.3.10) it is found that

$$I(f^* : f) = \frac{1}{2} \int \frac{|M(\lambda)|^2 (d\lambda)^2 (\lambda^2 + \beta^2)}{2\beta\, d\lambda} = \frac{1}{2} \int \frac{|M(\lambda)|^2}{2\beta} (\lambda^2 + \beta^2)\, d\lambda . \tag{2.7.5}$$

But from (2.7.4) using the properties of the Fourier transform

$$\int (m(t))^2\, dt = \int |M(\lambda)|^2\, d\lambda \tag{2.7.6}$$

$$m'(t) = \int \frac{e^{it\lambda}}{(2\pi)^{\frac{1}{2}}} (i\lambda) M(\lambda)\, d\lambda \tag{2.7.7}$$

$$\int (m'(t))^2\, dt = \int \lambda^2 |M(\lambda)|^2\, d\lambda , \tag{2.7.8}$$

hence

$$I(f^*:f) = \frac{1}{4}\int\left[\frac{(m'(t))^2}{\beta} + \beta(m(t))^2\right]dt \tag{2.7.9}$$

which is the same as Chapter IV (2.33) for $-\infty < t < \infty$ rather than $0 \le t \le T$ with $m(-\infty) = 0 = m(\infty)$.

The linear discriminant functional is determined from (2.3.9) and (2.3.3) as

$$\overline{U}(\lambda) = \frac{M(\lambda)\,d\lambda\,(\lambda^2+\beta^2)}{2\beta\,d\lambda} = \frac{(\lambda^2+\beta^2)M(\lambda)}{2\beta} = \int\frac{e^{-it\lambda}}{(2\pi)^{\frac{1}{2}}}\,u(t)\,dt \tag{2.7.10}$$

$$\begin{aligned} u(t) &= \int e^{it\lambda}\,\frac{(\lambda^2+\beta^2)}{(2\pi)^{\frac{1}{2}}\,2\beta}\,M(\lambda)\,d\lambda \\ &= -\frac{1}{2\beta}\,\frac{d^2}{dt^2}\,m(t) + \frac{\beta}{2}\,m(t) \end{aligned} \tag{2.7.11}$$

$$Z = \int u(t)X(t)\,dt = \frac{1}{2}\int X(t)\left[\beta\,m(t) - \frac{1}{\beta}\,m''(t)\right]dt \tag{2.7.12}$$

which agrees with (9.3.10) page 78 of Lilliefors (1964), provided the time is taken as $-\infty < t < \infty$ rather than $0 \le t \le T$.

<u>Example 2.8</u> See Example 2.4. Let us take the signal and noise case again in that we consider

$$X(t) = \int_{-\infty}^{\infty}\frac{e^{it\lambda}}{(2\pi)^{\frac{1}{2}}}\,dy(\lambda),\quad \operatorname{cov}(X(s),X(t)) = \int_{-\infty}^{\infty}\frac{e^{i\lambda(s-t)}}{2\pi}\,dF(\lambda) \tag{2.8.1}$$

where we take

$$E\,|\,dy\,(\lambda)\,|^2 \;=\; dF\,(\lambda) \;=\; \sigma^2\,d\lambda,\;\; \sigma\,(s-t) \;=\; \sigma^2\,\delta\,(s-t)\,. \tag{2.8.2}$$

If we take $Y(t)$ as in (2.4.4), where now

$$S\,(t) \;=\; \int_{-\infty}^{\infty} \frac{e^{it\lambda}}{(2\pi)^{\frac{1}{2}}}\; M\,(\lambda)\,d\lambda \tag{2.8.3}$$

then

$$I\,(f^*:f) \;=\; \int \frac{|\,M\,(\lambda)\,|^2\,(d\lambda)^2}{2dF\,(\lambda)} \;=\; \int \frac{|\,M\,(\lambda)\,|^2\,(d\lambda)^2}{2\sigma^2\,d\,\lambda} \tag{2.8.4}$$

$$=\; \int \frac{|\,M\,(\lambda)\,|^2\,d\lambda}{2\sigma^2}\,.$$

From (2.8.3) it is found that

$$\int\;(S\,(t)\,)^2\,dt \;=\; \int\;|\,M\,(\lambda)\,|^2\;d\lambda \tag{2.8.5}$$

and hence

$$I\,(f^*:f) \;=\; \frac{1}{2\sigma^2}\int\;(S\,(t)\,)^2\;dt\,. \tag{2.8.6}$$

For this case we define the associated generalized process (see (2.1.32) - (2.1.37)) by

$$y_n \;=\; \int\;X\,(t)\,a_n\,(t)\;dt \tag{2.8.7}$$

where $\{a_n\,(t)\,\}$ is a complete set of orthonormal functions,

$$\mathrm{Var}(y_n) = \int\int a_n(s)\, a_n(t)\, \sigma^2 \delta(s-t)\, ds dt = \sigma^2 , \tag{2.8.8}$$

$$\mathrm{Cov}(y_m y_n) = \int\int a_m(s)\, a_n(t)\, \sigma^2 \delta(s-t)\, ds dt = 0 , \tag{2.8.9}$$

$$d_n = \int S(t)\, a_n(t)\, dt . \tag{2.8.10}$$

The total energy in the signal is

$$E = \int (S(t))^2\, dt = \int |M(\lambda)|^2\, d\lambda \tag{2.8.11}$$

and the mean noise power per unit bandwidth is

$$N = f(\lambda) + f(-\lambda) = \sigma^2 + \sigma^2 = 2\sigma^2 . \tag{2.8.12}$$

Thus (2.8.6) may be written as

$$I(f^* : f) = E / N , \tag{2.8.13}$$

that is, we get again the signal to noise ratio. (See Kullback (1959), page 9.)

Note that if the noise in (2.8.1) is passed through a filter with transfer function

$$\Phi(\lambda) = \begin{cases} 1/(2W)^{\frac{1}{2}}, & -2\pi W \le \lambda \le 2\pi W \\ 0 & \text{elsewhere} \end{cases} \tag{2.8.14}$$

the output is given by

$$Z(t) = \int \frac{e^{it\lambda}}{(2\pi)^{\frac{1}{2}}}\, \Phi(\lambda)\, dy(\lambda) = \int_{-2\pi W}^{2\pi W} \frac{e^{it\lambda}}{(2W2\pi)^{\frac{1}{2}}}\, dy(\lambda) \tag{2.8.15}$$

and

$$\mathrm{Cov}\,(Z(s), Z(t)) = \sigma_Z(s-t) = \int \frac{e^{i\lambda(s-t)}}{2\pi} \,|\Phi(\lambda)|^2\, dF(\lambda) \tag{2.8.16}$$

$$= \int_{-2\pi W}^{2\pi W} \frac{e^{i\lambda(s-t)}}{2W} \, \frac{\sigma^2 d\lambda}{2\pi} = \sigma^2 \, \frac{\sin 2\pi W(s-t)}{2\pi W(s-t)} ,$$

with $dF_Z(\lambda) = \sigma^2 \, d\lambda / 4\pi W$.

The signal in (2.8.3) when passed through the filter (2.8.14) provides the output

$$m(t) = \int \frac{e^{it\lambda}}{(2\pi)^{\frac{1}{2}}} \, \Phi(\lambda) M(\lambda)\, d\lambda = \int_{-2\pi W}^{2\pi W} \frac{e^{it\lambda}}{(2W2\pi)^{\frac{1}{2}}} \, M(\lambda)\, d\lambda . \tag{2.8.17}$$

For the corresponding filtered hypotheses as in (2.4.4), that is,

$$\begin{aligned} H_2 &: Y(t) = Z(t) \\ H_1 &: Y(t) = Z(t) + m(t) \end{aligned} \tag{2.8.18}$$

$$I(1:2) = \tfrac{1}{2} \int \frac{|\phi(\lambda)|^2 |M(\lambda)|^2 d\lambda}{|\phi(\lambda)|^2 \dfrac{dF(\lambda)}{d\lambda}} = \frac{4\pi W}{2\sigma^2} \int_{-2\pi W}^{2\pi W} \frac{|M(\lambda)|^2}{2W} \, d\lambda \tag{2.8.19}$$

$$= \frac{2W}{2\sigma^2} \int (m(t))^2 \, dt$$

and indeed we arrive at the results in (2.4.25) - (2.4.35), remembering that $X(t)$ in (2.4.25) is $Z(t)$ in (2.8.15); $dy(\lambda)$ in (2.4.25) is $dy(\lambda)/(2W)^{\frac{1}{2}}$ in (2.8.15); $dF(\lambda)$ in (2.4.26) is $dF_Z(\lambda)$ in (2.8.16); $S(t)$ in (2.4.27) is $m(t)$ in (2.8.17); $d\theta(\lambda)$ in (2.4.27) is $M(\lambda)\, d\lambda/(2W)^{\frac{1}{2}}$ in (2.8.17).

<u>Example 2.9</u> Suppose the normal process $X(t)$ has the representation

$$X(t) = \int a(t,\lambda)\,dy(\lambda),\quad E|dy(\lambda)|^2 = dF(\lambda),\quad \sigma(s,t) = \int a(s,\lambda)\,\overline{a(t,\lambda)}\,dF(\lambda) \tag{2.9.1}$$

where $\overline{a(t,\lambda)}$ is the complex conjugate of $a(t,\lambda)$. The moment generating functional of $X(t)$ is

$$M = E\{exp \int u(t)X(t)\,dt\} = E\{exp \int\int u(t)\,a(t,\lambda)\,dy(\lambda)\,dt\} \tag{2.9.2}$$

$$= E\{exp \int a(\lambda)\,dy(\lambda)\},\quad a(\lambda) = \int u(t)\,a(t,\lambda)\,dt\,.$$

It is found that with $E(X(t)) = m(t) = \int a(t,\lambda)\,d\nu(\lambda)$

$$ln\,M = \int a(\lambda)\,d\nu(\lambda) + \frac{1}{2}\int |a(\lambda)|^2\,dF(\lambda) \tag{2.9.3}$$

and as in the preceding examples for

$$Z = \int u(t)X(t)\,dt = \int a(\lambda)\,dy(\lambda) \tag{2.9.4}$$

$$E(Ze^Z)/E(e^Z) = \int a(\lambda)\,d\nu(\lambda) + \int |a(\lambda)|^2\,dF(\lambda)\,, \tag{2.9.5}$$

$$\int a(\lambda)\,d\theta(\lambda) = \int a(\lambda)\,d\nu(\lambda) + \int |a(\lambda)|^2\,dF(\lambda) \tag{2.9.6}$$

$$I(f^* : f) = \int a(\lambda)\,d\theta(\lambda) - \int a(\lambda)\,d\nu(\lambda) - \frac{1}{2}\int |a(\lambda)|^2\,dF(\lambda) \tag{2.9.7}$$

$$= \frac{1}{2} \int |a(\lambda)|^2 dF(\lambda) .$$

From (2.9.6) it is seen that

$$\int a(\lambda)[\, d(\theta(\lambda) - \mathrm{v}(\lambda)) - \overline{a(\lambda)}\, dF(\lambda)] = 0 \tag{2.9.8}$$

that is

$$d(\theta(\lambda) - \mathrm{v}(\lambda)) = \overline{a(\lambda)}\, dF(\lambda) \tag{2.9.9}$$

hence

$$I(f^* : f) = \int \frac{|d(\theta(\lambda) - \mathrm{v}(\lambda))|^2}{2\, dF(\lambda)} \tag{2.9.10}$$

and

$$Z = \int a(\lambda)\, dy(\lambda) = \int \frac{\overline{d(\theta(\lambda) - \mathrm{v}(\lambda))}\, dy(\lambda)}{dF(\lambda)} . \tag{2.9.11}$$

Note that

$$\sigma_Z^2 = \int |a(\lambda)|^2 dF(\lambda) = \int a(\lambda)\, d(\theta(\lambda) - \mathrm{v}(\lambda)) \tag{2.9.12}$$

but

$$\left| \int a(\lambda) \cdot \; d(\theta(\lambda) - \mathrm{v}(\lambda)) \right|^2 \le \int |a(\lambda)|^2 dF(\lambda) \int \frac{|d(\theta(\lambda) - \mathrm{v}(\lambda))|^2}{dF(\lambda)} \tag{2.9.13}$$

hence, if $\int a(\lambda)\, d(\theta(\lambda) - \nu(\lambda))$ is finite, but $\int \frac{|d(\theta(\lambda) - \nu(\lambda))|^2}{dF(\lambda)} = 2I(f^* : f)$ $= \infty$, then since

$$I(f^* : f) = \frac{|\int a(\lambda)\, d(\theta(\lambda) - \nu(\lambda))|^2}{2\sigma_Z^2}$$

it is noted that $\sigma_Z^2 = 0$ and there is perfect discrimination.

3. Poisson Process

The examples in the preceding section dealt with normal or Gaussian processes with a common covariance structure under the hypotheses. We shall now examine another type of process.

Consider a continuous time Poisson process on the interval $0 \le t \le T$. This process is defined as a series of events for which the occurrence of any single event during the time interval $(t, t + h)$ is independent of the number of previous occurrences and its probability is $\lambda(t)\, h + o(h)$ where $\lambda(t)$ is a function of t and $o(h)$ implies that for a function $g(h) = o(h)$, $\lim g(h)/h = 0$ for $h \to 0$.

Consider the hypotheses

$$H_1 : \lambda(t) = \lambda_1(t), \qquad H_2 : \lambda(t) = \lambda_2(t)\,.$$

The increment $\Delta x(t) = x(t + \Delta t) - x(t)$ has a Poisson distribution with

$$E_i\, \Delta x(t) = \Delta\mu_i(t) = \mu_i(t + \Delta t) - \mu_i(t)\,,\ i = 1, 2 \tag{3.1}$$

where

$$\mu_i(t) = \int_o^t \lambda_i(z)\,dz\,. \tag{3.2}$$

Assume $\mu_i(T) < \infty, i = 1, 2$ so that $\mu_1 \sim \mu_2$. This condition implies that $\{t \in [0, T]: \lambda_1(t) = 0\} = \{t \in [0, T] : \lambda_2(t) \geq 0\}$ see Brown (1971). For the Poisson variable $\Delta x(t)$ we have (Kullback (1959, p. 144))

$$I(1:2\,;\Delta x(t)) = \Delta\mu_1(t)\,ln\,\frac{\Delta\mu_1(t)}{\Delta\mu_2(t)} + \Delta\mu_2(t) - \Delta\mu_1(t)\,. \tag{3.3}$$

Since the Poisson process has independent increments

$$\begin{aligned} I(1:2\,;x(t), 0 \leq t \leq T) &= \sup \sum I(1:2\,;\,\Delta x(t)) \\ &= \lim_{\Delta t \to o} \sum (\Delta\mu_1(t)\,ln\,\frac{\Delta\mu_1(t)}{\Delta\mu_2(t)} + \Delta\mu_2(t) - \Delta\mu_1(t)) \\ &= \int_0^T (d\mu_1(t)\,ln\,\frac{d\mu_1(t)}{d\mu_2(t)} + d\mu_2(t) - d\mu_1(t)). \end{aligned} \tag{3.4}$$

The minimum discrimination information estimates are

$$\hat{I}(*:2) = \int_0^T (dx(t)\,ln\,\frac{dx(t)}{d\mu_2(t)} + d\mu_2(t) - dx(t)) \tag{3.5}$$

$$\hat{I}(*:1) = \int_0^T (dx(t)\,ln\,\frac{dx(t)}{d\mu_1(t)} + d\mu_1(t) - dx(t)) \tag{3.6}$$

and the log-likelihood ratio is

$$\hat{I}(*:2) - \hat{I}(*:1) = \int_0^T (dx(t)\,ln\,\frac{d\mu_1(t)}{d\mu_2(t)} + d\mu_2(t) - d\mu_1(t))\,. \tag{3.7}$$

(See Duttweiler and Kailath (1973, p. 26).)

Since $dx(t) = 1$ for $t = t_j$ when an event occurs and $dx(t) = 0$ otherwise, the log-likelihood ratio, except for an additive constant is

$$\sum_{1}^{n} ln \frac{\lambda_1(t_j)}{\lambda_2(t_j)} \tag{3.8}$$

where n is a random variable indicating the number of occurrences in the interval $[0, T]$. Equation (3.8) is equivalent to the result in Grenander (1950) by another approach. If $\lambda_i(t) = \lambda_i$, a constant

$$I(1:2\ ;\ x(t),\ 0 \leq t \leq T) = T(\lambda_1\ ln \frac{\lambda_1}{\lambda_2} + \lambda_2 - \lambda_1)\ . \tag{3.9}$$

VI: CANONICAL CORRELATION

We shall now consider an extension to stationary Gaussian processes of the discussion relative to canonical correlations in Kullback (1959), pp. 200 - 204. We shall assume that the spectral distribution function is absolutely continuous and write

$$\frac{dF}{d\lambda} = f(\lambda)$$

where $f(\lambda)$ is the spectral density. We shall first examine the values of the spectral density as the eigenvalues of the covariance operator.

1. Spectral Densities as Eigenvalues.

Let us consider the representation of the stationary process in the form

$$x(t) = \frac{1}{(2\pi)^{1/2}} \int e^{it\lambda} dy(\lambda)$$

and its covariance function

$$\sigma(s-t) = \frac{1}{2\pi} \int_{-\infty}^{\infty} e^{i\lambda(s-t)} dF(\lambda) = \frac{1}{2\pi} \int_{-\infty}^{\infty} e^{i\lambda(s-t)} f(\lambda)\, d\lambda\,,$$

$$E \mid dy(\lambda) \mid^2 = dF(\lambda)\,. \tag{1.1}$$

The right-hand-side of (1.1) is a generalization of the spectral representation in (1.4) of chapter V with the eigenfunction $\frac{1}{(2\pi)^{1/2}} e^{i\lambda s}$ and eigenvalue $f(\lambda)$. Indeed, we have from (1.1)

$$\int_{-\infty}^{\infty} \sigma(s-t)\; e^{i\lambda' t}\, dt = \frac{1}{2\pi} \int\!\!\int e^{i\lambda' t + i\lambda s - i\lambda t} f(\lambda)\, d\lambda\, dt$$

$$= \int e^{i\lambda s} f(\lambda)\, \delta(\lambda' - \lambda)\, d\lambda = e^{i\lambda' s} f(\lambda') \tag{1.2}$$

that is,

$$\int_{-\infty}^{\infty} \sigma(s-t)\, \frac{e^{i\lambda' t}}{(2\pi)^{1/2}}\, dt = f(\lambda')\, \frac{e^{i\lambda' s}}{(2\pi)^{1/2}} \tag{1.3}$$

corresponding to (1.8) of chapter V. We also note from (1.1) that for $f(\lambda) = 1,\ \sigma(s-t) = \delta(s-t)$, and

$$\int_{-\infty}^{\infty} \frac{e^{it\lambda}}{(2\pi)^{1/2}} \frac{e^{-it\lambda'}}{(2\pi)^{1/2}}\, dt = \delta(\lambda - \lambda'). \tag{1.4}$$

The eigenvalues of the inverse covariance operator $\bar{\sigma}$ are $1/f(\lambda)$, that is

$$\bar{\sigma}(t-u) = \int_{-\infty}^{\infty} \frac{e^{i\lambda(t-u)}}{2\pi f(\lambda)}\, d\lambda \tag{1.5}$$

since

$$\int_{-\infty}^{\infty} \sigma(s-t)\, \bar{\sigma}(t-u)\, dt = \frac{1}{(2\pi)^2} \int\!\!\int\!\!\int e^{i\lambda(s-t)} f(\lambda)\, \frac{e^{i\lambda'(t-u)}}{f(\lambda')}\, d\lambda\, d\lambda'\, dt$$

$$= \frac{1}{2\pi} \int\!\!\int \frac{f(\lambda)}{f(\lambda')}\, e^{i\lambda s - i\lambda' u}\, \delta(\lambda' - \lambda)\, d\lambda\, d\lambda'$$

$$= \frac{1}{2\pi} \int e^{i\lambda(s-u)}\, d\lambda = \delta(s-u). \tag{1.6}$$

For the analogue of the matrix product $\Sigma_1 \Sigma_2^{-1}$ we have

$$\sigma_1 \overline{\sigma}_2 (s-u) = \int \sigma_1 (s-t) \overline{\sigma}_2 (t-u)\, dt$$

$$= \frac{1}{(2\pi)^2} \iiint e^{i\lambda(s-t)} f_1(\lambda) \cdot \frac{e^{i\lambda'(t-u)}}{f_2(\lambda')} d\lambda d\lambda' dt$$

$$= \frac{1}{2\pi} \iint \frac{f_1(\lambda)}{f_2(\lambda')} e^{i\lambda s - i\lambda' u} \delta(\lambda' - \lambda)\, d\lambda d\lambda'$$

$$= \frac{1}{2\pi} \int e^{i\lambda(s-u)} \frac{f_1(\lambda)}{f_2(\lambda)} d\lambda \tag{1.7}$$

that is, the eigenvalues of the operator $\sigma_1 \overline{\sigma}_2$ are the values $f_1(\lambda)/f_2(\lambda)$.

For the analogue of the matrix product Σ^2 we have

$$\sigma\sigma(s-u) = \int_{-\infty}^{\infty} \sigma(s-t)\sigma(t-u)\, dt$$

$$= \frac{1}{(2\pi)^2} \iiint e^{i\lambda(s-t)} f(\lambda) \cdot e^{i\lambda'(t-u)} f(\lambda')\, d\lambda d\lambda' dt$$

$$= \frac{1}{(2\pi)} \iint f(\lambda) f(\lambda') e^{i\lambda s - i\lambda' u} \delta(\lambda' - \lambda)\, d\lambda d\lambda'$$

$$= \frac{1}{2\pi} \int e^{i\lambda(s-u)} (f(\lambda))^2 d\lambda , \tag{1.8}$$

that is, the eigenvalues of the operator $\sigma\sigma = \sigma^2$ are values $(f(\lambda))^2$, with similar results for other powers.

Corresponding to the trace of a matrix, that is, the sum of the diagonal elements, or the sum of the eigenvalues, we have

$$\mathrm{tr}\sigma = \lim_{T\to\infty} \int_{-T}^{T} \sigma(t-t)\, dt = \lim_{T\to\infty} 2T\sigma(0))$$

$$= \lim_{T \to \infty} \frac{2T}{2\pi} \int f(\lambda)\, d\lambda. \tag{1.9}$$

We recall that the determinant of a matrix is the product of the eigenvalues, so that we may write

$$ln\,|\mathbf{\Sigma}| = \sum_{i=1}^{n} ln\,\lambda_i = \text{tr}\; ln\;\mathbf{\Sigma}. \tag{1.10}$$

Since

$$ln\;\boldsymbol{\sigma}\,(s-t) = \frac{1}{2\pi} \int e^{i\lambda(s-t)}\, ln\; f(\lambda)\, d\lambda \tag{1.11}$$

we have that

$$\text{tr}\; ln\;\boldsymbol{\sigma} = \lim_{T \to \infty} \int_{-T}^{T} ln\;\sigma\;(t-t)\; dt = \lim_{T \to \infty} \frac{2T}{2\pi} \int ln\; f(\lambda)\, d\lambda\,. \tag{1.12}$$

2. Discrimination Information - Different Covariance Functions .

We shall now derive the value of the discrimination information rate for the case of a stationary Gaussian process with the same mean value but different covariance functions under the two hypotheses. We shall arrive at the expression already used in (2.56) of chapter IV.

Let x(t) be a stationary Gaussian process with zero mean and covariances

$$H_1 :\; Cov_1\,(x(s), x(t)) = \sigma_1\,(s-t) = \frac{1}{2\pi} \int e^{i\lambda(s-t)}\, f_1(\lambda)\, d\lambda\,,$$

$$H_2 : Cov_2(x(s), x(t)) = \sigma_2(s-t) = \frac{1}{2\pi}\int e^{i\lambda(s-t)} f_2(\lambda)\, d\lambda . \tag{2.1}$$

For the case of a multivariate normal vector with zero mean but different covariance matrices under the hypotheses we have from Kullback (1959), p. 190, p. 195

$$I(1:2) = \frac{1}{2}\left(ln\ \frac{|\Sigma_2|}{|\Sigma_1|} + \mathrm{tr}\ \Sigma_1 (\Sigma_2^{-1} - \Sigma_1^{-1})\right)$$

$$= \frac{1}{2}\ \mathrm{tr}\ \left(-\ ln\ \Sigma_1\ \Sigma_2^{-1} + \Sigma_1 (\Sigma_2^{-1} - \Sigma_1^{-1})\right)$$

$$= \frac{1}{2}\sum_{i=1}^{k} (-\ ln\ \lambda_i + \lambda_i - 1) \tag{2.2}$$

where the λ_i are the eigenvalues of $\Sigma_1\ \Sigma_2^{-1}$.

For the stationary Gaussian process the discrimination information value corresponding to the middle term of (2.2) is

$$I(1:2\ ;\ x(t)) = \frac{1}{2}\ \mathrm{tr}\ (-\ ln\ \sigma_1\overline{\sigma}_2 + \sigma_1\overline{\sigma}_2 - \sigma_1\overline{\sigma}_1) . \tag{2.3}$$

From the results in section 1 on the spectral densities as eigenvalues we have

$$-\ \mathrm{tr}\ ln\ \sigma_1\ \overline{\sigma}_2 = -\frac{2T}{2\pi}\int ln\ \frac{f_1(\lambda)}{f_2(\lambda)}\ d\lambda \tag{2.4}$$

$$\mathrm{tr}\ (\sigma_1\sigma_2 - \sigma_1\sigma_1) = \frac{2T}{2\pi}\int \left(\frac{f_1(\lambda)}{f_2(\lambda)} - 1\right) d\lambda \tag{2.5}$$

and hence the rate of discrimination information is

$$I = \lim_{T\to\infty} \frac{I\ (1:2\ ;x(t),\ -T \le t \le T)}{2T}$$

$$= \frac{1}{4\pi} \int \left(\frac{f_1(\lambda)}{f_2(\lambda)} - 1 - \ln \frac{f_1(\lambda)}{f_2(\lambda)} \right) d\lambda \tag{2.6}$$

with the conditions given in (2.87) and (2.88) of Chapter IV. The expression in (2.6) is of course the analogue of the last version in (2.2).

For a different approach to the problem, let us define the random variable (linear discriminant function)

$$Z_k = \frac{1}{(2T)^{1/2}} \int_{-T}^{T} x(t)\, e^{-it\lambda_k}\, dt \tag{2.7}$$

for T large. The variance of Z_k is

$$Var\ Z_k = \frac{1}{2T} \int\int \sigma(s-t)\, e^{-i\lambda_k(s-t)}\, dsdt \tag{2.8}$$

or using (1.3) we have for T very large and smoothly varying power spectrum

$$Var\ Z_k = \frac{1}{2T} \int f(\lambda_k)\, ds = f(\lambda_k). \tag{2.9}$$

We then have for the normal random variable Z_k

$$I(1:2;Z_k) = \frac{1}{2} \left(\frac{f_1(\lambda_k)}{f_2(\lambda_k)} - 1 - \ln \frac{f_1(\lambda_k)}{f_2(\lambda_k)} \right). \tag{2.10}$$

Since the Z_k are independent random variables and we saw in Example 2.4 of Chapter V, that for processes limited to the band (0, W) cycles per second and $0 \le t \le T$ there were $n = 2WT$ independent variables or $n = \Lambda T/\pi$ (frequency in radians is equal to 2π times frequency in cycles per second) that is,

$$\frac{\Lambda}{n} = \frac{\pi}{T} \quad or \quad \Delta\lambda = \frac{\pi}{T} . \tag{2.11}$$

Thus, the discrimination information in the stationary process is the sum of the discrimination information values over the narrow band filters, that is

$$I(1:2\,;x(t),\, -T \le t \le T) = \frac{T}{\pi}\frac{1}{2}\sum_{k}\left(\frac{f_1(\lambda_k)}{f_2(\lambda_k)} - 1 - ln\,\frac{f_1(\lambda_k)}{f_2(\lambda_k)}\right)\Delta\lambda \tag{2.12}$$

or

$$\bar{I} = \lim_{T\to\infty}\frac{I(1:2\,;x(t),\,-T\le t\le T)}{2T} = \frac{1}{4\pi}\int\left(\frac{f_1(\lambda)}{f_2(\lambda)} - 1 - ln\,\frac{f_1(\lambda)}{f_2(\lambda)}\right)d\lambda . \tag{2.13}$$

3. Stationarily Related Processes

Let us consider the stationary processes $x(t), y(t)$ with the representation

$$x(t) = \frac{1}{(2\pi)^{1/2}}\int e^{it\lambda}\,d\eta_x(\lambda),\quad y(t) = \frac{1}{(2\pi)^{1/2}}\int e^{it\lambda}\,d\eta_y(\lambda) \tag{3.1}$$

with covariance functions (* denotes the complex conjugate)

$$\mathrm{cov}\,(x(s)\,,x^*(t)) = \sigma_{xx}(s-t) = \frac{1}{2\pi}\int e^{i\lambda(s-t)}\,dF_{xx}(\lambda)$$

$$= \frac{1}{2\pi}\int e^{i\lambda(s-t)} f_{xx}(\lambda)\,d\lambda$$

$$\operatorname{cov}(y(s), y^*(t)) = \sigma_{yy}(s-t) = \frac{1}{2\pi}\int e^{i\lambda(s-t)} dF_{yy}(\lambda)$$

$$= \frac{1}{2\pi}\int e^{i\lambda(s-t)} f_{yy}(\lambda)\, d\lambda$$

$$E \mid d\eta_x(\lambda) \mid^2 = dF_{xx}(\lambda),\ E \mid d\eta_y(\lambda) \mid^2 = dF_{yy}(\lambda)\,. \tag{3.2}$$

We also define the cross-covariance functions

$$\operatorname{cov}(x(s), y^*(t)) = \sigma_{xy}(s-t) = \frac{1}{2\pi}\int e^{i\lambda(s-t)} dF_{xy}(\lambda)$$

$$= \frac{1}{2\pi}\int e^{i\lambda(s-t)} f_{xy}(\lambda)\, d\lambda$$

$$\sigma_{yx}(s-t) = \sigma^*_{xy}(s-t),\ E(d\eta_x(\lambda)\, d\eta^*_y(\lambda)) = dF_{xy}(\lambda) = dF^*_{yx}(\lambda) \tag{3.3}$$

where $F_{xy}(\lambda)$ is the cross-spectral distribution function and $f_{xy}(\lambda)$ the cross-spectral density of the stationarily related radom processes $x(t)$ and $y(t)$.

We now define the random process $z(s)$ as

$$z(s) = y(s) - \int\int \sigma_{yx}(s-u)\, \bar{\sigma}_{xx}(u-t)\, x(t)\, dt\, du \tag{3.4}$$

where $\bar{\sigma}_{xx}(u-t)$ is the inverse covariance function, that is,

$$\int \sigma_{xx}(s-t)\, \bar{\sigma}_{xx}(t-u)\, dt = \delta(s-u)\,. \tag{3.5}$$

From (3.4) it is found that

$$= \operatorname{cov} \Big(y(s) - \int\!\!\int \sigma_{yx}(s-u)\, \bar{\sigma}_{xx}(u-v)\, x(v)\, dv\, du\,,$$

$$y^*(t) - \int\!\!\int \sigma^*_{yx}(t-u')\, \bar{\sigma}_{xx}(u'-v')\, x^*(v')\, dv'\, du' \Big)$$

$$= \sigma_{yy}(s-t) - \int\!\!\int \sigma_{yx}(s-v)\, \bar{\sigma}_{xx}(v-u)\, \sigma_{xy}(u-t)\, dv\, du$$

$$- \int\!\!\int \sigma_{yx}(s-u)\, \bar{\sigma}_{xx}(u-v)\, \sigma_{xy}(v-t)\, dv\, du$$

$$+ \int\!\!\int\!\!\int\!\!\int \sigma_{yx}(s-u) \bar{\sigma}_{xx}(u-v) \sigma_{xx}(v-v') \bar{\sigma}_{xx}(u'-v') \sigma^*_{yx}(t-u')\, dv du dv' du'$$

$$= \sigma_{yy}(s-t) - \int\!\!\int \sigma_{yx}(s-v)\, \bar{\sigma}_{xx}(v-u)\, \sigma_{xy}(u-t)\, du\, dv \tag{3.6}$$

since the four-fold integral in (3.6) may be written as

$$\int\!\!\int\!\!\int \sigma_{yx}(s-u)\, \bar{\sigma}_{xx}(u-v) \sigma^*_{yx}(t-u')\; \delta\,(u'-v)\, dv\, du\, du'$$

$$= \int\!\!\int \sigma_{yx}(s-u)\, \bar{\sigma}_{xx}(u-v) \sigma_{xy}(v-t)\; du\, dv\,. \tag{3.7}$$

We get the spectral density of $z(s)$ as follows, using the spectral and cross-spectral densities of $x(s)$, $y(s)$ in the final relation in (3.6) ,

$$\sigma_{zz}(s-t) = \frac{1}{2\pi} \int e^{i\lambda(s-t)} f_{yy}(\lambda)\, d\lambda$$

$$- \int\!\!\int\!\!\int\!\!\int\!\!\int \frac{1}{2\pi} e^{i\lambda(s-v)} f_{yx}(\lambda)\, d\lambda\, \frac{e^{i\lambda'(v-u)} d\lambda'}{2\pi f_{xx}(\lambda')}\, \frac{1}{2\pi}\, e^{i\lambda''(u-t)} f_{xy}(\lambda'')\, d\lambda''\, du\, dv \tag{3.8}$$

and the five-fold integral may be written as

$$\int\!\!\int\!\!\int\!\!\int \frac{1}{2\pi} e^{i\lambda(s-v)} f_{yx}(\lambda)\, d\lambda\, \frac{e^{i\lambda' v} d\lambda'}{2\pi f_{xx}(\lambda')}\, e^{-i\lambda'' t} f_{xy}(\lambda'')\; \delta\,(\lambda''-\lambda')\, d\lambda'' dv$$

$$= \int\int\int \frac{1}{2\pi} e^{i\lambda(s-v)} f_{yx}(\lambda)\, d\lambda \, \frac{e^{i\lambda' v} d\lambda'}{2\pi f_{xx}(\lambda')} e^{-i\lambda' t} f_{xy}(\lambda')\, dv$$

$$= \int\int \frac{1}{2\pi} e^{i\lambda s} f_{yx}(\lambda)\, d\lambda \, \frac{d\lambda'}{f_{xx}(\lambda')} e^{-i\lambda' t} f_{xy}(\lambda')\, \delta(\lambda' - \lambda)$$

$$= \frac{1}{2\pi} \int e^{i\lambda(s-t)} \frac{f_{yx}(\lambda) f_{xy}(\lambda)}{f_{xx}(\lambda)}\, d\lambda \tag{3.9}$$

so that

$$\sigma_{zz}(s-t) = \frac{1}{2\pi} \int e^{i\lambda(s-t)} \{ f_{yy}(\lambda) - \frac{f_{yx}(\lambda) f_{xy}(\lambda)}{f_{xx}(\lambda)} \}\, d\lambda \tag{3.10}$$

that is

$$f_{zz}(\lambda) = f_{yy}(\lambda) - f_{yx}(\lambda) f_{xy}(\lambda) / f_{xx}(\lambda)\,. \tag{3.11}$$

We also get the spectral representation of $z(s)$ from (3.4), as follows

$$z(s) = \frac{1}{(2\pi)^{1/2}} \int e^{i\lambda s} d\eta_z(\lambda) = \frac{1}{(2\pi)^{1/2}} \int e^{i\lambda s} d\eta_y(\lambda)$$

$$- \int\int\int\int\int \frac{1}{2\pi} e^{i\lambda(s-u)} f_{yx}(\lambda)\, d\lambda \, \frac{e^{i\lambda'(u-t)}\, d\lambda'}{2\pi f_{xx}(\lambda')} \frac{1}{(2\pi)^{1/2}} e^{i\lambda'' t} d\eta_x(\lambda'')\, dt du \tag{3.12}$$

and the five-fold integral may be written as

$$\int\int\int\int \frac{1}{2\pi} e^{i\lambda(s-u)} f_{yx}(\lambda)\, d\lambda \frac{e^{i\lambda' u} d\lambda'}{f_{xx}(\lambda')} \frac{1}{(2\pi)^{1/2}} d\eta_x(\lambda'')\ \delta\ (\lambda'' - \lambda')\, du$$

$$= \int\int\int \frac{1}{2\pi} e^{i\lambda(s-u)} f_{yx}(\lambda)\, d\lambda \frac{e^{i\lambda' u} d\lambda'}{f_{xx}(\lambda')} \frac{1}{(2\pi)^{1/2}} d\eta_x(\lambda')\, du$$

$$= \int\int e^{i\lambda s} f_{yx}(\lambda)\, d\lambda \frac{d\lambda'}{f_{xx}(\lambda')} \frac{1}{(2\pi)^{1/2}} d\eta_x(\lambda')\ \delta\ (\lambda' - \lambda)$$

$$= \frac{1}{(2\pi)^{1/2}} \int e^{i\lambda s} \frac{f_{yx}(\lambda)}{f_{xx}(\lambda)} d\eta_x(\lambda) \tag{3.13}$$

so that

$$z(s) = \frac{1}{(2\pi)^{1/2}} \int e^{i\lambda s} d\eta_z(\lambda) = \frac{1}{(2\pi)^{1/2}} \int e^{i\lambda s} \left(d\eta_y(\lambda) - \frac{f_{yx}(\lambda)}{f_{xx}(\lambda)} d\eta_x(\lambda)\right) \tag{3.14}$$

that is

$$d\eta_z(\lambda) = d\eta_y(\lambda) - \frac{f_{yx}(\lambda)}{f_{xx}(\lambda)} d\eta_x(\lambda)\,. \tag{3.15}$$

Note that

$$E\,(d\eta_z(\lambda)\, d\eta_x^*(\lambda)) = E\,(d\eta_y(\lambda)\, d\eta_x^*(\lambda)) - \frac{f_{yx}(\lambda)}{f_{xx}(\lambda)} E\,|\,d\eta_x(\lambda)\,|^2$$

$$= dF_{yx}(\lambda) - \frac{f_{yx}(\lambda)}{f_{xx}(\lambda)} dF_{xx}(\lambda) = 0 \tag{3.16}$$

and we also have from (3.4) that

$$\operatorname{cov}(z(s), x^*(v)) = \operatorname{cov}(y(s) . x^*(v))$$

$$- \int\int \sigma_{yx}(s-u)\, \bar{\sigma}_{xx}(u-t)\, \sigma_{xx}(t-v)\, dtdu$$

$$= \sigma_{yx}(s-v) - \int \sigma_{yx}(s-u)\ \delta\ (u-v)\ du$$

$$= \sigma_{yx}(s-v) - \sigma_{yx}(s-v) = 0\ . \tag{3.17}$$

As a check, we have from (3.15)

$$E \mid d\eta_z(\lambda) \mid^2 = E \mid d\eta_y(\lambda) - \frac{f_{yx}(\lambda)}{f_{xx}(\lambda)}\, d\eta_x(\lambda) \mid^2$$

$$= E \mid d\eta_y(\lambda) \mid^2 - \frac{f_{xy}(\lambda)}{f_{xx}(\lambda)}\, E\,(d\eta_y(\lambda)\, d\eta_x^*(\lambda))$$

$$- \frac{f_{yx}(\lambda)}{f_{xx}(\lambda)}\, E\,(d\eta_x(\lambda)\, d\eta_y^*(\lambda)) + \frac{f_{yx}(\lambda)\, f_{xy}(\lambda)}{f_{xx}^2(\lambda)}\, E \mid d\eta_x(\lambda) \mid^2$$

$$= dF_{yy}(\lambda) - \frac{f_{xy}(\lambda)}{f_{xx}(\lambda)}\, dF_{yx}(\lambda) - \frac{f_{yx}(\lambda)}{f_{xx}(\lambda)}\, dF_{xy}(\lambda)$$

$$+ \frac{f_{yx}(\lambda)\, f_{xy}(\lambda)}{f_{xx}^2(\lambda)}\, dF_{xx}(\lambda) \tag{3.18}$$

which reduces to (3.11). We may also introduce the correlation between $d\eta_x(\lambda)$ and $d\eta_y(\lambda)$ by

$$\rho^2(\lambda) = \frac{(E\,(d\eta_x(\lambda)\,d\eta_y^*(\lambda)))^2}{E\,|\,d\eta_x(\lambda)\,|^2 E\,|\,d\eta_y(\lambda)|^2} = \frac{dF_{xy}(\lambda)\,dF_{yx}(\lambda)}{dF_{xx}(\lambda)\,dF_{yy}(\lambda)}$$

$$= \frac{f_{xy}(\lambda)\,f_{yx}(\lambda)}{f_{xx}(\lambda)\,f_{yy}(\lambda)} \tag{3.19}$$

and note that (3.11) may also be written as

$$f_{zz}(\lambda) = f_{yy}(\lambda)\,(1-\rho^2(\lambda)). \tag{3.20}$$

4. Gaussian Processes

We now assume that $y(s)$ and $x(t)$ in (3.4) are Gaussian processes, so that $z(s)$ is also Gaussian. We consider the hypotheses

$$H_1 : z(s) = y(s) - \int\int \sigma_{yx}(s-u)\,\bar{\sigma}_{xx}(u-t)\,x(t)\,dt\,du$$

$$H_2 : z(s) = y(s) \tag{4.1}$$

in other words, that $\sigma_{yx}(s-u) = 0$ under H_2, or $y(t)$ and $x(t)$ are independent under H_2. In terms of the spectral representations, the hypotheses in (4.1) are equivalent to

$$H_1 : d\eta_z(\lambda) = d\eta_y(\lambda) - \frac{dF_{yx}(\lambda)}{dF_{xx}(\lambda)}\,d\eta_x(\lambda)$$

$$H_2 : d\eta_z(\lambda) = d\eta_y(\lambda). \tag{4.2}$$

We observe that the right-hand side of (4.1) is the analogue of the vector expression

$$x_{2.1} = x_2 - \Sigma_{21}\Sigma_{11}^{-1}x_1 \tag{4.3}$$

on p. 201 of Kullback (1959) with the vector $\mathbf{x}_2$ corresponding to $y(s)$ and the vector $\mathbf{x}_1$ to $x(t)$ and the vector residual $\mathbf{x}_{2.1}$ to $z(s)$. The analogue of the covariance matrix of $\mathbf{x}_{2.1}$, that is $\Sigma_{22.1}$ is the operator σ_{zz} defined in (3.6). In the corresponding vector case in Kullback (1959), (7.4) p. 201 we found that

$$I(1:2) = \frac{1}{2}\, ln \frac{|\Sigma_{22}|}{|\Sigma_{22.1}|} \tag{4.4}$$

which we can also write as (Kullback (1959, p. 203)),

$$\frac{1}{2}\, ln \frac{|\Sigma_{22}|}{|\Sigma_{22.1}|} = -\frac{1}{2}\, ln\,|\,\Sigma_{22.1}\Sigma_{22}^{-1}| = -\frac{1}{2}\, tr\; ln\; \Sigma_{22.1}\Sigma_{22}^{-1}$$
$$= -\frac{1}{2}\sum_i ln\;(1-\rho_i^2) \tag{4.5}$$

where the ρ_i^2 are the squared canonical correlations. For the case of the stochastic processes as defined in (4.1) we must then find the value of

$$-\frac{1}{2}\;\mathrm{tr}\; ln\;\sigma_{zz}\,\bar{\sigma}_{yy}\,. \tag{4.6}$$

Using (1.7) , (1.12) , (3.10) it is found that

$$\sigma_{zz}\,\bar{\sigma}_{yy}\,(s-t) = \int \sigma_{zz}\,(s-u)\,\bar{\sigma}_{yy}\,(u-t)\,du$$

$$= \frac{1}{2\pi}\int e^{i\lambda(s-t)}\left(1 - \frac{f_{yx}(\lambda)\,f_{xy}(\lambda)}{f_{xx}(\lambda)\,f_{yy}(\lambda)}\right)d\lambda$$

$$= \frac{1}{2\pi}\int e^{i\lambda(s-t)}\,(1-\rho^2(\lambda))\,d\lambda \tag{4.7}$$

where $\rho^2(\lambda)$ was defined in (3.19) . Hence

$$\operatorname{tr}\, ln\ \sigma_{zz}\,\bar{\sigma}_{yy} = \frac{2T}{2\pi}\int ln\ (1-\rho^2(\lambda))\,d\lambda \tag{4.8}$$

and (see Pinsker (1964 , p. 181))

$$\lim_{T\to\infty}\frac{I(1:2\,;z(t),\,-T\le t\le T)}{2T} = -\frac{1}{4\pi}\int ln\ (1-\rho^2(\lambda))\,d\lambda. \tag{4.9}$$

An alternative approach in terms of the formulation in (4.2) follows (see (2.64) in Chapter IV).

$$I\,(1:2\,;d\eta_z(\lambda)) = E\,(I\,(1:2\,;d\eta_z(\lambda)\mid d\eta_x(\lambda)))$$

$$= \frac{1}{2}E\left(\left|\frac{dF_{yx}(\lambda)}{dF_{xx}(\lambda)}\,d\eta_x(\lambda)\right|^2 / dF_{yy}(\lambda)\right)$$

$$+\frac{1}{2}\left(\frac{dF_{yy}(\lambda)-\dfrac{dF_{yx}(\lambda)\,dF_{xy}(\lambda)}{dF_{xx}(\lambda)}}{dF_{yy}(\lambda)} - 1 - ln\,\frac{dF_{yy}(\lambda)-\dfrac{dF_{yx}(\lambda)\,dF_{xy}(\lambda)}{dF_{xx}(\lambda)}}{dF_{yy}(\lambda)}\right)$$

$$= \frac{1}{2}\frac{dF_{yx}(\lambda)\,dF_{xy}(\lambda)\,dF_{xx}(\lambda)}{(dF_{xx}(\lambda))^2\,dF_{yy}(\lambda)} - \frac{1}{2}\frac{dF_{yx}(\lambda)dF_{xy}(\lambda)}{dF_{yy}(\lambda)\,dF_{xx}(\lambda)}$$

$$- \frac{1}{2} \; ln \; (1 \; - \; \frac{dF_{yx}(\lambda)\, dF_{xy}(\lambda)}{dF_{yy}(\lambda)\, dF_{xx}(\lambda)})$$

$$= - \frac{1}{2} \; ln \; (1 \; - \; \rho^2 \,(\lambda)) \, . \tag{4.10}$$

Thus using (2.11) , the information over the spectrum is

$$- \frac{T}{2\pi} \sum \; ln \; (1 - \rho^2 \,(\lambda))\, \Delta\lambda \tag{4.11}$$

and

$$\lim_{T \to \infty} \frac{I(1:2\,;z\,(t)\,,\, -T \leq t \leq T)}{2T} = - \frac{1}{4\pi} \int ln \; (1 - \rho^{\,2} \,(\lambda))\; d\,\lambda . \tag{4.12}$$

As a particular case of interest let us assume that

$$\int \sigma_{yx}(s-u)\,\bar{\sigma}_{xx}(u-t)\; du \; = \; \delta\,(s-t) \tag{4.13}$$

or equivalently

$$\frac{1}{2\pi} \int e^{i\,\lambda\,(s-t)} \, \frac{f_{yx}(\lambda)}{f_{xx}(\lambda)} \; d\,\lambda \; = \; \frac{1}{2\pi} \int e^{i\,\lambda\,(s-t)} \; d\,\lambda \tag{4.14}$$

then (3.4) becomes

$$z\,(s) \; = \; y\,(s) \; - \int x\,(t) \;\, \delta \;\, (s-t) \; dt \; = \; y\,(s) \; - \; x\,(s) \tag{4.15}$$

and equivalently (3.15) and (3.11) become respectively

$$d\eta_z\,(\lambda) \; = \; d\eta_y\,(\lambda) \; - \; d\eta_x\,(\lambda)\;, \tag{4.16}$$

$$f_{zz}(\lambda) = f_{yy}(\lambda) - f_{xx}(\lambda) . \tag{4.17}$$

In this case (4.9) yields

$$\lim_{T\to\infty} \frac{I(1:2; z(t), -T\leq t\leq T)}{2T} = -\frac{1}{4\pi}\int ln\,(1-\frac{f_{xx}(\lambda)}{f_{yy}(\lambda)})\,d\lambda$$

$$= \frac{1}{4\pi}\int ln\,\frac{f_{yy}(\lambda)}{f_{zz}(\lambda)}\,d\lambda = \frac{1}{4\pi}\int ln\,\frac{f_{zz}(\lambda)+f_{xx}(\lambda)}{f_{zz}(\lambda)}\,d\lambda$$

$$= \frac{1}{4\pi}\int ln\,(1+\frac{f_{xx}(\lambda)}{f_{zz}(\lambda)})\,d\lambda. \tag{4.18}$$

If $x(t)$, $y(t)$, $z(t)$ are respectively the transmitted signal, the received signal, and the noise then the integrand in (4.18) is the logarithm of 1 plus the signal to noise ratio at the frequency λ in the spectrum.

In particular, if we take

$$\begin{aligned} f_{xx}(\lambda) &= \frac{\sigma_x^2 d\lambda}{4\pi W},\ f_{zz}(\lambda) = \frac{\sigma_n^2 d\lambda}{4\pi W},\ -2\pi W\leq\lambda\leq 2\pi W \\ &= 0 \quad , \qquad\quad = 0 \quad , \quad \text{elsewhere} \end{aligned} \tag{4.19}$$

we find from (4.18)

$$\lim_{T\to\infty} \frac{I\,(1:2;\,z(t),\ -T\leq t\leq T)}{2T}$$

$$= \frac{1}{4\pi}\int_{-2\pi W}^{2\pi W} ln\,(1+\frac{\sigma_x^2}{\sigma_n^2})\,d\lambda = W\,ln\,(1+\frac{\sigma_x^2}{\sigma_n^2})$$

$$= W\,ln\,(1+\frac{S}{N}), \tag{4.20}$$

that is, the same information rate found earlier for this special case.

REFERENCES

Anorina, L.A. (1970) "On the Question of the Equivalence of two Gaussian Measures " (Russian). *Stochastic Processes and Related Problems*. Part 1 (Russian) pp. 3-25 Izdat. "Fan" Uzbek SSR, Tashkent. MR 43#2768.

Borpujari, A.S. (1971) "Asymptotic Methods for Estimating the Mean of N Stationary Time Series". Ph. D. Dissertation, The George Washington University.

Borpujari, A.S. (1977) "An empirical Bayes Approach for Estimating the Mean of N Stationary Time Series". *J. Amer. Statist. Ass.*, 72, 397-402.

Borpujari, A.S.
Kullback, S. (1983) "Application of Minimum Discrimination Information in Stationary Time Series". *Journal of Combinatorics, Information & System Sciences*, 8 No. 1, 61-72.

Brown, M. (1971) "Discrimination of Poisson Processes". *Ann. Math. Statist*, 42, 773-776.

Capon, J. (1965) "Hilbert Space Methods for Detection Theory and Pattern Recognition". *IEEE Trans. Inform. Theory*, IT-11, 247-259.

Carslaw, H.S. (1930) *Fourier's Series and Integrals*. Dover Publication, Inc.

Chiang Tse-Pei (1958) "A Note on the Definition of Information". *Theory of Probability and Its Applications* (English translation) 3, 93-97.

Chung, K.L. (1968) *A Course in Probability Theory*, Harcourt, Brace and World, Inc. N. Y.

Dobrushin, R.L. (1959) "General Formulation of Shannon's Main Theorem in Information Theory". (Russian) *Usp. Mat. Nauk.*, 14, No. 6, 3-104. English translation, *A.M. S. Translations, Series 2*, 33(1963), 323-438.

Dobrushin, R.L. (1960) "Passage to the Limit Under the Information and Entropy Signs". *Theory of Probability and Its Applications*. (English translation) 5, 25-32.

Doob, J.L. (1953) *Stochastic Processes*, Wiley, New York

Duttweiler, D.L.
Kailath, T. (1973) "RKHS Approach to Detection and Estimation Problems - Part IV: Non-Gaussian Detection". *IEEE Trans. Inform. Theory*, IT-19, 19-28

Gelfand, I.M.
Kolmogorov, A.N.
Yaglom, A.M. (1956) "On the General Definition of the Amount of Information". (Russian) *Doklady Akad, Nauk, SSSR.*, 111, 745-748. German translation: *Arbeiten zur Informations theorie II*, VEB Deutscher Verlag der Wissenshaften, Berlin 1958, 57-60.

References (Continued)

Gelfand, I.M. Yaglom, A.M. (1957) "Calculation of the Amount of Information About a Random Function Contained in Another Such Function". (Russian) *Usp. Mat. Nauk.*, 12, No. 1(73), 3-52. English translation: *A.M.S. Translations, Series 2*, 12 (1959), 199-246.

Ghurye, S.G. (1968) "Information and Sufficient Sub-fields". *Annals of Math. Statist.* 39, 2056-66.

Giasu, S. (1968) *Applications of Information Theory*, (Romanian), Bucarest 36-40.

Gokhale, D.V. Kullback, S. (1978) *The Information in Contingency Tables*, Marcel Dekker Inc., New York

Grenander, U. I. (1950) "Stochastic Processes and Statistical Inference". *Arkiv för Matematik*, 1, 195-277.

Grenander, U. I. (1967) *Eight Lectures on Statistical Inference in Stochastic Processes*, Tech. Report No. 2, May 1967, Division of Applied Mathematics, Brown Univ.

Hájek, J. (1958). "A Property of J-Divergence of Marginal Probability Distributions". *Czechoslovak Math. J.* 8 (83), 460-3.

Halmos, P.R. (1950) *Measure Theory*, D. Van. Nostrand Co., Inc., Princeton, N.J.

Helstrom, C.W. (1960) *Statistical Theory of Signal Detection*. Pergamon Press, New York, 2nd ed. 1968.

Ikeda, S. (1960) "A Remark on the Convergence of Kullback-Leibler's Mean Information". *Ann. Inst. Stat. Math.* (Tokyo)., 12, 81-88.

Kadota, T.T. (1965) "Optimum Reception of Binary Sure and Gaussian Signals". *Bell Syst. Tech. J.*, 44, 1621-1658.

Kallianpur, G. (1960) "On the Amount of Information Contained in a Sigma-Field". *Contributions to Probability and Statistics - Essays in Honor of Harold Hotelling*. Stanford University Press. 265-73.

Ku, H.H., Varner, R.N. Kullback, S. (1971) "On the Analysis of Multidimensional Contingency Table". *J. Amer. Statist. Ass.*, 66, 55-64.

Kullback, S. (1959) *Information Theory and Statistics*. Wiley, N.Y; 1968 Edition Dover Publications, Inc., N.Y.; 1978, Peter Smith Pub. Inc., Magnolia MA; 1967, (Russian), Moscow.

Kullback, S. (1966) "An Information-Theoretic Derivation of Certain Limit Relations for a Stationary Markov Chain". *J. SIAM Control*, 4, No. 3, 454-459.

References (Continued)

Kullback, S. (1967a) "An Extension of an Information-Theoretic Derivation of Certain Limit Relations for a Markov Chain". *J. SIAM Control*, 5, No. 1, 51-53.

Kullback, S. (1967) "A Lower Bound for Discrimination Information In Terms of Variation". *IEEE Transactions on Information Theory*, IT-13,126-7.

Kullback, S. (1968) "On the Convergence of Discrimination Information". *IEEE Transactions on Information Theory*, IT-14, 765-6.

Kullback, S. (1970) "Minimum Discrimination Information Estimation and Application-Invited Paper Presented to Sixteenth Conference on the Design of Experiments in Army Research, Development and Testing", U.S.A. Logistics Management Center, Ft. Lee, VA. 21 October 1970. *Proceedings of the Conference ARO-D Report*, 71-3, 1-38.

Kullback, S. Keegel, J.C. (1984) "Categorical Data Problems Using Information Theoretic Approach". *Handbook of Statistics*, Vol. 4, *Nonparametric Methods*, P.R. Krishnaiah and P.K. Sen, eds., North Holland Publishing Company, Amsterdam, The Netherlands.

Kullback, S. Khairat, M.A. (1966) "A Note on Minimum Discrimination Information". *Annals of Math. Statist.* 37,279-80.

Kullback, S. Thall, P.F. (1977) "An Information Theoretic Proof of the Integral Representation Theorem". *Journal of Combinatorics, Information and System Sciences*, 2,97-103.

Lilliefors, H. W. (1964) "An Information Theoretic Application of Linear Discriminant Functionals to Stochastic Processes". Ph.D. Dissertation, The Graduate Council, The George Washington University.

Loève, M. (1955) *Probability Theory*. D. Van Nostrand Co., Inc., Second Edition, 1960; Third Edition, 1963.

Melnick, E. L. (1973) "An Application of the Minimum Discrimination Information Estimate for Detecting Gaussian Signal". *IEEE Trans. Inform. Theory*, IT-19 No. 3,367-369.

Melnick, E.L. Kullback, S. (1973) "An Application of Minimum Discrimination Information Estimate to Compute Log-Likelihood Ratios". *Journal of Applied Probability*, 10,469-474.

References (Continued)

Neveu, J. (1965) *Mathematical Foundations of the Calculus of Probabilities*. Holden-Day, Inc. (Translated from the French 1964 edition by A. Feinstein.)

Parzen, E. (1982) "Maximum Entropy Interpretation of Autoregressive Spectral Densities". *Statistics and Probability Letters*, 1, 7-11.

Peirce, B.O. (1929) *A Short Table of Integrals*, Third revised editor, Ginn and Co.

Pérez, A. (1957) "Notions generalisees d'incertitude d'entropie et d'information du point de vue de la theorie de martingales". *Transactions of the first Prague conference on information theory statistical decision functions, random processes*. November 28 to 30 1956. Publishing house of the Czechoslovak Academy of Sciences, Prague, 183-208.

Pinsker, M.S. (1964) *Information and Information Stability of Random Variables and Processes*, Holden-Day, Inc. (translated by A. Feinstein from the Russian published in Moscow 1960).

Rényi, A. (1962) *Wahrscheinlichkeitsrechnung*, DVW, Berlin, page 298.

Rosenblatt, M. (1962) *Random Processes*, Oxford University Press, N.Y.

Rosenblatt, M. (1963) (Editor) *Time Series Analysis*, Wiley, New York.

Rozanov, Yu. A. (1963) *Stationary random Processes*, (Russian) Moscow.

Shumway, R.H. (1965) "An Information Theoretic Approach to the Problem of Discriminating Between Two Stochastic Processes". Ph.D. dissertation, The Graduate Council, The George Washington University.

Simon, G.A. (1971) "Information Distances and Exponential Families, With Applications to Contingency Tables". Dep. Statist., Stanford University, Stanford, CA,Tech. Rep. 32, Nov. 26, 1971.

Striebel, C.T. (1959) "Densities for Stochastic Processes". *Annals of Math. Statist.*, 30,559-67.

Yaglom, A.M. (1962) *Theory of Stationary Random Functions*, Prentice-Hall Translated by R. A. Silverman from the Russian.

Yaglom, A.M. (1963) "On the Equivalence or Perpendicularity of Two Gaussian Probability Measures in Function Space". *Proceedings of the Symposium on Time Series Analysis*, 1962, *Brown University*, Wiley, New York, 327-346.

INDEX